GOOD HONEST TALES

150 YEARS OF BATEMANS BREWERY

ADAM CARTWRIGHT

AMBERLEY

First published 2024

Amberley Publishing
The Hill, Stroud,
Gloucestershire, GL5 4EP

www.amberley-books.com

ISBN: 978 1 3981 1769 3 (print)
ISBN: 978 1 3981 1770 9 (ebook)

British Library Cataloguing in Publication Data.
A catalogue record for this book is available from the British Library.

Typeset in 10pt on 13pt Celeste.
Typesetting by SJmagic DESIGN SERVICES, India.
Printed in the UK.

Contents

Acknowledgements

I have not included specific references and footnotes in the text, but this book could not have been written without access to Batemans' own archives, some of which date back to the earliest years of the business. These include correspondence written by George and Harry Bateman (the Letter Books); share, property and mortgage registers; a range of business ledgers; annual financial statements; newspaper and magazine cuttings; and photographs, to name but a few. I also made extensive use of the British Newspaper Archive website, especially the *Boston Guardian, Stamford Mercury* and *Lincolnshire Standard.* Records of other brewers were helpful too, especially Soames of Spalding (Norfolk Record Office) and Mowbrays of Grantham (Lincolnshire Archives).

I would also like to thank the following for their willingness to be interviewed, and the time and help they so generously gave me: Stuart Bateman, Jaclyn Bateman, Stuart Bell, Martin Cullimore, Sandra Raymond and Neil Hutchinson at Batemans. Many thanks as well to Ken Redmore and Andrew Walker for kindly reviewing and commenting on drafts of the book, and my wife Michelle for proofreading. I am also grateful to several photographers (credited in the text) who freely gave permission for me to use their work. Finally, Connor Stait and Alison Flowers at Amberley were very supportive throughout this project.

Opposite: This map of Batemans' pub locations was produced for a promotional poster in around 1987, showing how their pubs were clustered in the east of Lincolnshire for many years. The inset shows the twelve pubs in Boston – by far their biggest presence. At this stage the tied estate stood at 105, and the firm were just about to branch out of the county, although significant pub sales were also soon to take place. The map was produced for a promotional poster. (Batemans)

Map of Batemans' Pub Locations, *c*. 1987

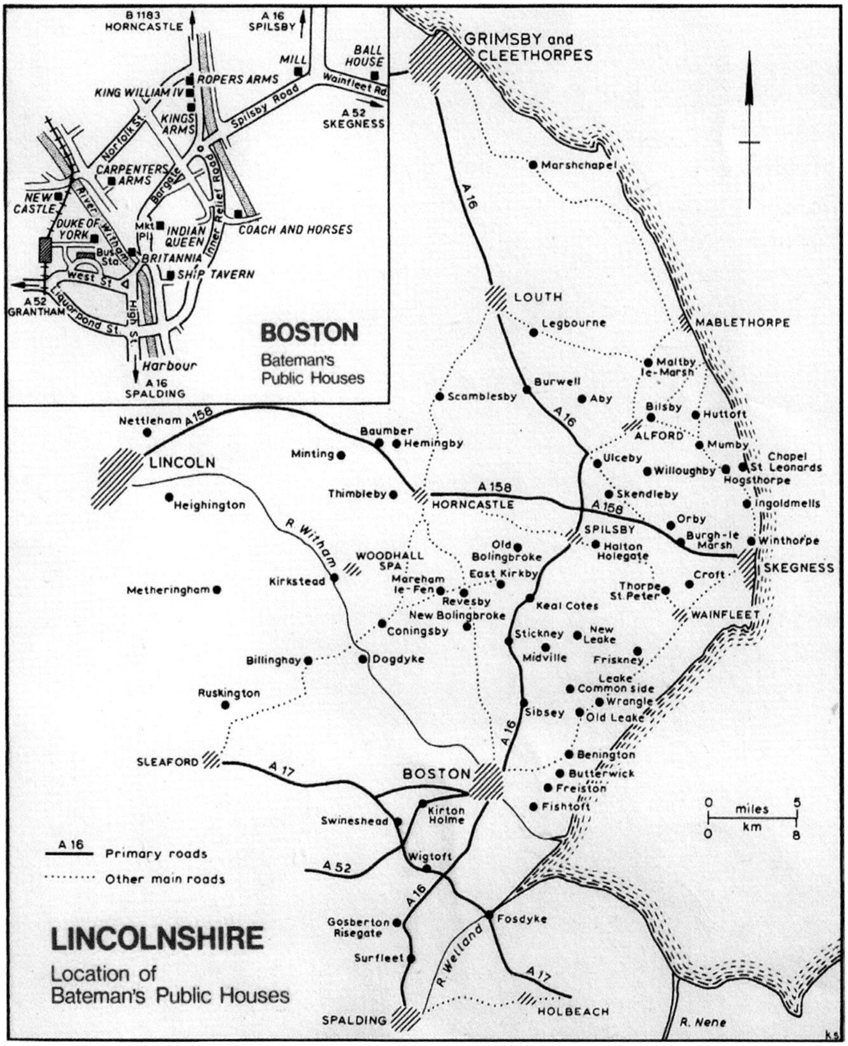

Introduction

It's a beautiful summer's morning in 1874, just before eight o'clock. The heat is already rising with the promise of another blistering day as a local farmer hauls on the reins of his cart to direct the old horse across Salem Bridge on his approach to the small market town of Wainfleet.

The cart rattles over the Steeping River, the horse's hooves clattering as he trots past the first houses of the town. Somebody is already out with a ladder, cutting back the ivy on the beautiful house on his left with its cast-iron balconies.

Despite the problems of farming, George Bateman is a happy and cheery man. He and his wife Susannah are hard workers, but George has just turned thirty and he's ambitious. The Batemans have been farmers for as long as any of them can remember, but he would like to do other things as well. Or perhaps instead.

His thoughts are interrupted by the sight of his friend Edwin Crow, one of the local brewers, who has just popped into view as he turns his own cart out of Church Lane. Edwin's got a few beer barrels from his brewery on the back and clearly he's about to start deliveries to local farmers. Probably not a brewing day, then, George guesses. He greets his friend cheerily with a wave of his arm. 'Mayut!' ('Mate', pronounced in the local accent, is the traditional form of greeting around here.)

George would like a chat with Edwin; he needs a new barrel of beer, maybe two, as the harvest season is upon them, and farmhands can get very thirsty. So he slows his cart right down. Edwin has done the same thing; best to make the most of this unplanned meeting. He returns George's greeting. 'Now then, me ode mayut. Susannah all right?'

'She's fine thanks, Edwin. 'Ow are you doing, after all that fuss and bother wi' your lad down Skeggy beach?'

'He's safely ovver in Boston now, and I 'ope that's all behind us. It's just this damned cough I can't seem to get rid of. Trade's 'olding up well though. Which reminds me ...'

He leans across the reins with a conspiratorial glint in his eye, and checks there's nobody around who could possibly hear. 'George, 'ow'd you like to buy the brewery off me?'

George is momentarily lost for words, but recovers quickly. He plays for time.

'Give ovver, Edwin, what do I know about brewing? And, more to the point, 'ow would we agree a price?'

Edwin senses the opportunity. 'Them's just details, George. I can teach you brewing, and I can fix you up with old Blind George to help. After a bit of practice, you'll be fine. And as for the money, well, you could pay me back ovver time.' He pauses. 'I'm not getting any younger, you know. But the price needs to be right, mind you.'

George smiles to himself. He senses Edwin is desperate to sell. 'All right, leave it with me and I'll have a word with Susannah, see what she thinks. Don't speak to anyone else. I'll let you have my reply by Saturday.'

Whether it really took place like that, well, who knows, but the story that's been handed down is that Edwin Crow met George Bateman by chance one day on Wainfleet's Salem Bridge and offered to sell his brewery. So maybe it wasn't much different from my take on the conversation.

Fast forward almost 150 years later, and I'm enjoying yet another classic pint of Batemans XXXB, but I'm not in a Batemans pub, or anywhere in the UK for that matter. I'm nearly

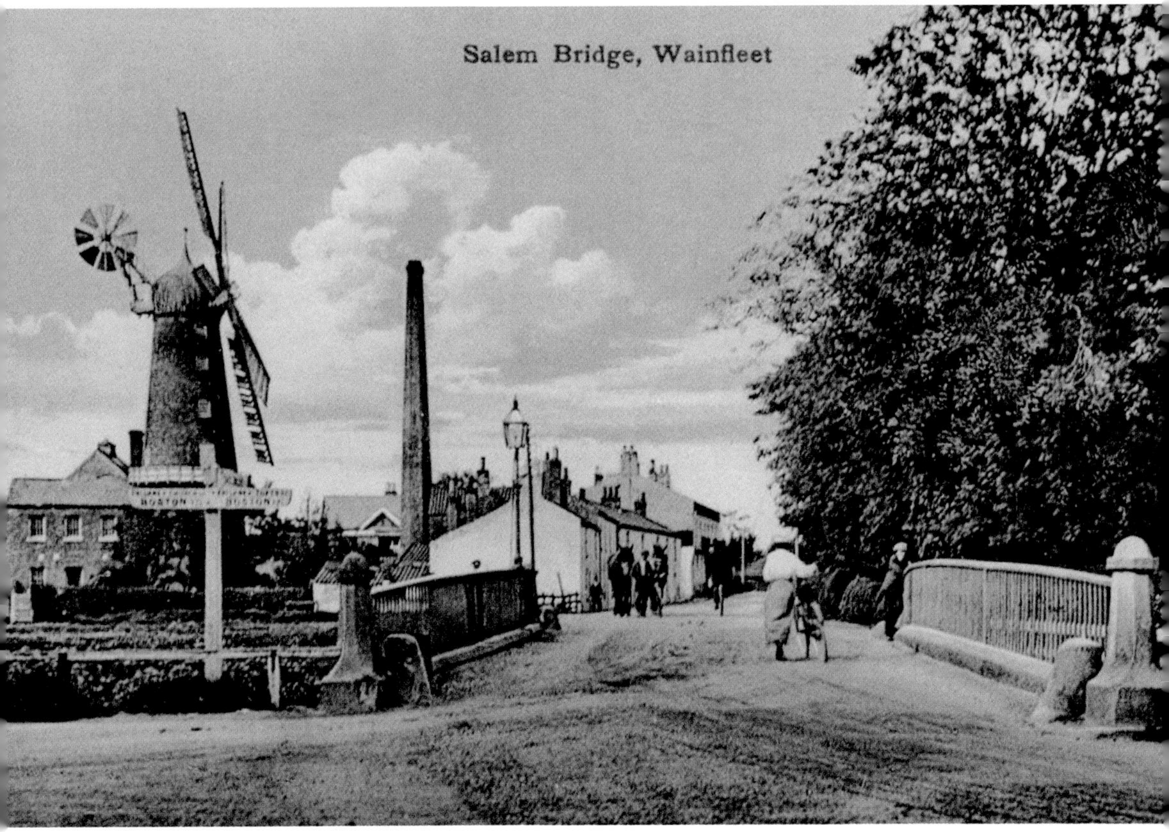

Conversation Piece
The bridge over the Steeping River at Wainfleet, where the imagined conversation between George Bateman and Edwin Crow took place. Two horses are being led south out of the town towards the cyclist wheeling her cycle over the bridge. The fashions and the fact that the mill appears to be in working order suggests that the image dates from around 1900. The signpost indicates that Boston is equally accessible in two directions!

4,000 miles from Lincolnshire, yet the beer is in great condition, although the pub I'm in is very slightly rocking about. And no, I've not had one too many.

I'm on a cruise ship, approaching Antigua in the Caribbean. Finding XXXB on hand pump in the ship's 'pub' is a nice surprise, but also typical of the enterprising approach of this proudly independent brewer as it approaches its significant anniversary. I suspect the first George Bateman would have been astonished at the eventual reach and range of his firm's beers; the generations who succeeded him, well, maybe not.

How Batemans progressed from very small beginnings to a modern twenty-first-century business is a fascinating tale to unravel. Often fraught with challenge and difficulty, at times so severe that it seemed impossible to survive and brew another day, the business has reached its 150th anniversary through a combination of determination and enterprise on the part of its owners, tenants, customers and staff – with just a smattering of good fortune and successful risk-taking along the way.

Join me to explore the story of five generations and a unique brewing business.

1

Before 1874 – A Horde of Crows

The Crow family were long-established farmers in and around Wainfleet, long enough for their family name to be celebrated in the name of a bridge over the Steeping River and a nearby country lane. William Crow established a brewery around 1810 on a site in Church Lane, Wainfleet, close to the river, the location no doubt chosen to provide easy access to a water supply for his fledgling business. As an indication of his status as a respected and reliable local figure, William accumulated a number of other local posts and occupations: he became an auctioneer, a tax collector and an Overseer of the Poor, in charge of administering parish relief (a very primitive and largely ineffective form of social security). His farming interests eventually extended to over 400 acres, based at Chain Bridge Farm to the south of the town. William handed over the brewery to his son Edwin when he was just twenty-two. Edwin expanded the operation, adding a small malt kiln in 1845 for which he engaged a specialist maltster. The brewery was eventually capable of producing around 250 gallons/1,100 litres of 'fine sparkling ales' per brewing session – it was what we would now call a microbrewery. However, there is no evidence that the Crows ever owned or leased any public houses, suggesting perhaps a lack of available cash for expansion.

Edwin was a busy man, for in addition to the business – which he named South End Brewery – he was an auctioneer and in time took over his father's farming interests. He had also branched out into valuation, insurance and estate agency, which took up more and more of his time as he expanded his trade beyond Wainfleet to the much bigger local town of Boston, 18 miles to the south-west. He owned property in and around Wainfleet, including the town's main hotel, the Woolpack. This multiplicity of roles was not unusual for the time, but it may have led to a lack of focus and supervision at the brewery and could have contributed to an appalling accident there in April 1858. Edwin had employed local innkeeper Benjamin Overton to brew for him, who was about to take over as landlord at the Red Lion in Wainfleet, making it quite likely that he was filling in as a favour to Crow. Overton was leaning over the mash tun, which was full of boiling liquor, when he overbalanced and fell in, being so badly burned that he later died of his injuries. Crow shortly afterwards improved working arrangements at the brewery and engaged 'an experienced practical Brewer'.

In 1869 Edwin Crow passed the brewing and malting businesses over to his only son, George. As well as wanting to concentrate on his farming, valuation and auction businesses, he'd just become a director of the fledgling Wainfleet and Firsby railway company, which in a few years would link the town to the national rail network. Following the death of his first wife, he was courting Caroline Longden, a young woman twenty-one years his junior and not much older than his son; they married the next year. With Edwin's attentions elsewhere, George Crow took over the brewery at the age of just twenty-two (the same age at which his father had been given the brewery), but within a year he was charged with drunkenness and fined, an unfortunate omen for what was to come. It's possible that the change of management was dictated by the state of Edwin's health, who may already have been suffering from the effects of the neurological illness that was to kill him a few years later. Edwin presided over his last auction in March 1873, advertising his services for a few more months as a valuer.

It was from this point that the fortunes of the Crow family and their South End brewery started to spiral out of control. The maltings side of the business probably ended in 1872 when the malthouse and cottage were advertised to let. It's also likely that part of the brewery site was needed when the Wainfleet branch railway was extended on to Skegness in July 1873, as the new line curved sharply away from the station to head north-east, cutting across the brewery's land. The other parts of the family business fell away after Edwin's retirement; his son George seems to have presided over just one auction, of a couple of unpromising properties, despite resorting to advertising his services in the local papers. The pressures all came to a head on an August day in 1874 when George Crow, many sheets to the wind, harnessed his horse to a cart and forced it to gallop the 5 miles to Skegness. By 'sundry thrashings' he whipped the valuable animal on across the sands and into the advancing tide, where it fell, probably from exhaustion, and drowned. Crow himself was only narrowly rescued from the same fate, luckily escaping with just a few cuts. The drunken spectacle Crow had made of himself in front of many horrified spectators, and his appalling treatment of the horse, could not be ignored: he was charged first with drunkenness and then animal cruelty. Although the magistrates considered a jail sentence, he was instead fined heavily – perhaps they were unaware of his earlier conviction.

These were clearly the actions of a very troubled man, perhaps one unable to cope with the stresses of a business thrust into his hands by an increasingly ill father. Having placed himself so firmly in the public spotlight, George Crow had blown whatever credibility he might have had as a businessman, and the reaction of his wife Elizabeth, whose father was the Superintendent of Police at Boston, can only be imagined. Allowing some weeks for the dust to settle, his father tried a damage limitation exercise, starting with this announcement in the county's leading newspaper on 23 October 1874: 'EDWIN CROW begs to announce that he has resumed the Business of BREWER, and respectfully solicits the patronage of his old friends and supporters.' However, this revival was always going to be brief, because in the same edition he advertised the brewery for rent, describing it as a family business established sixty-four years ago. He also invited his 'old friends' to dine at the brewery on 24 September 1874. Shortly after that, Edwin Crow let the brewery to George Bateman, and it is tempting to think that the two of them might have finalised a deal over dinner that Saturday. We shall probably never know, but a week later Edwin announced an auction of his son's quarter share in the family's properties in the town, placing the sale in the hands of a former competitor.

The Crows soon faded from the scene. The brewery's founder, William, had died in 1868, having reached eighty years old. Edwin, in increasingly poor health, followed eight years later aged only fifty. His wife Caroline returned to London, her birthplace, with some of their children; married for just six years, she would survive her husband by another forty-six years. The errant George had left Wainfleet at the end of 1874, almost certainly to escape his notoriety there. Some of his furniture, guns and other possessions were auctioned off on New Year's Eve that year, the sale document tersely stating 'he is leaving'. Six years later, George died aged just thirty-three at the Blue Lion beerhouse in Stanbow Lane, Boston, one of the town's poorest and roughest streets. The pub was also a lodging house, a notorious haunt of prostitutes and thieves, and it seems very likely that George was living there. The cause of death was recorded as pneumonia, which can be caused by alcohol abuse; it's possible that George drank himself to death. The informant was his sister Salome, rather than his wife, which suggests they had separated. Salome was living a few streets away from the pub, but with her own problems; within eighteen months of her brother's death, her husband deserted her and their two children. Although the Crow family history does not make for easy reading, the difficult circumstances of key individuals and the eventual failure of their brewery was key to the start of the Batemans story.

2

Beginnings, 1874–1900

The Bateman family's roots in Lincolnshire go back hundreds of years, at least as far as the sixteenth century and perhaps even earlier. Not surprisingly for such a rural county, many of them were farmers. Our story starts with John Bateman, who farmed a few acres of fertile marshland soil a couple of miles from the sea near the village of Friskney. The farmhouse was situated between two tight bends on the main road from Boston to Skegness. It was, and remains, an exposed and mainly treeless spot: in the winter at the mercy of howling gales from the North Sea, in the summer peaceful and tranquil, and at all times the vast Lincolnshire skies sweep overhead. John had taken on the farm sometime in the 1830s. He and his wife Elizabeth believed in large families, like many farmers, hoping that at least some of them would help out on the farm and eventually take it over. They had ten children, but the one who is of particular interest to us is George Bateman, born in 1844 when his parents were already in their early forties. It was he who founded the Bateman brewing business.

John Bateman's death in July 1866, at the age of sixty-seven, prompted the sale of the farm, which was advertised for sale shortly afterwards by Edwin Crow, in his auctioneer role. Not only was there the farm at Friskney – which in all comprised 21 acres – John also owned a further 14 acres a few miles away, a cottage with 2 more acres of land in Wainfleet, as well as sheep, horses and other livestock. John's son, Thomas, bought most of the Friskney farm. That left 4 acres, which were sold to a farmer named Pickering who sub-leased them to the young George Bateman. However, this was not enough for George, who within a few years had taken a twenty-year lease on a 'desirable farm' in Friskney covering a further 48 acres, more land in fact than his father had farmed, with nearly 100 animals – sheep, cattle, horses and ponies.

How George Bateman – still only in his twenties – managed to expand his farming business so quickly is not clear. He could have been left some money by his father, but it can't have been much – he had three brothers and six sisters, and his father's farming interests were fairly modest. But his wife Susannah (née Simpson), whom he married in 1867, had her own money. Her father, who died nine years before the marriage, had been a local farmer and coal merchant, and the family also owned a farm in the southern Fens at Algarkirk, which Susannah inherited. Perhaps it was a combination of her wealth and support, and the shrewd business brain that George would display in his brewing business; he certainly knew how to drive a good bargain.

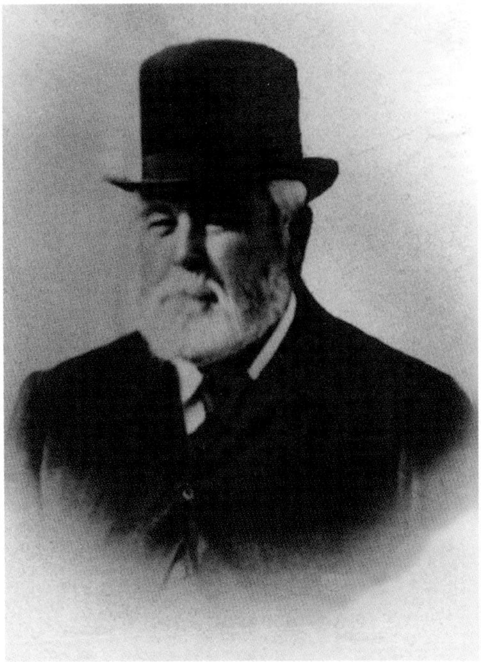

Dignified Founder
George Bateman, in a very formal portrait probably recorded by a local photographer around the first decade of the twentieth century, when he was in his sixties. He wears a smartly trimmed full beard and top hat at a time when fashions were starting to turn away from both styles, although the curved top of his hat reflects a trend towards bowlers. (Batemans)

What is very clear, though, is that early in 1875 George Bateman advertised the lease of his farm, and the sale of his animals and farm equipment. Although he had not completely finished with farming, he had been sufficiently impressed by his conversation with Edwin Crow to make a leap into the unknown in the brewing industry. George may also have had concerns about the future of farming; if he did, they were well founded. The relative economic prosperity of the 1860s and early 1870s was beginning to downturn; whilst the harvest of 1874 had been one of the best for some time, the following years saw poor weather conditions producing later, lower-yielding harvests. Whilst this was still not good news for a brewer, it was much worse for farmers. Whether he sensed this coming or not, George had again chosen wisely.

Having agreed to rent Edwin Crow's South End brewery in late 1874, George bought the brewing equipment outright for £505 and 10*s*. Crow almost certainly provided George with details of all his existing customers, which would have provided some reliable income from the outset. A year later he bought the lease of the South End brewery premises for £800, and also the present brewery house, Salem House, which would have been even more expensive – if we only knew how much he paid for it. It's a vast Georgian brick building dating from around 1825, with a distinctive cast-iron balcony to the first floor. Initially, George rented out much of the house, using some of the rooms as offices, but he may well have been thinking of moving the brewery over there. For the first few years, however, George seems to have traded from the old South End premises. The unusual name of the house, and that of nearby Salem Bridge, both derive from Sailholme, a nearby but long-gone parcel of land, perhaps almost an island, in a loop of the Steeping River.

In my imagined conversation with Crow, George claims not to know anything about brewing – or at least he might have wanted to give Edwin Crow that impression. The

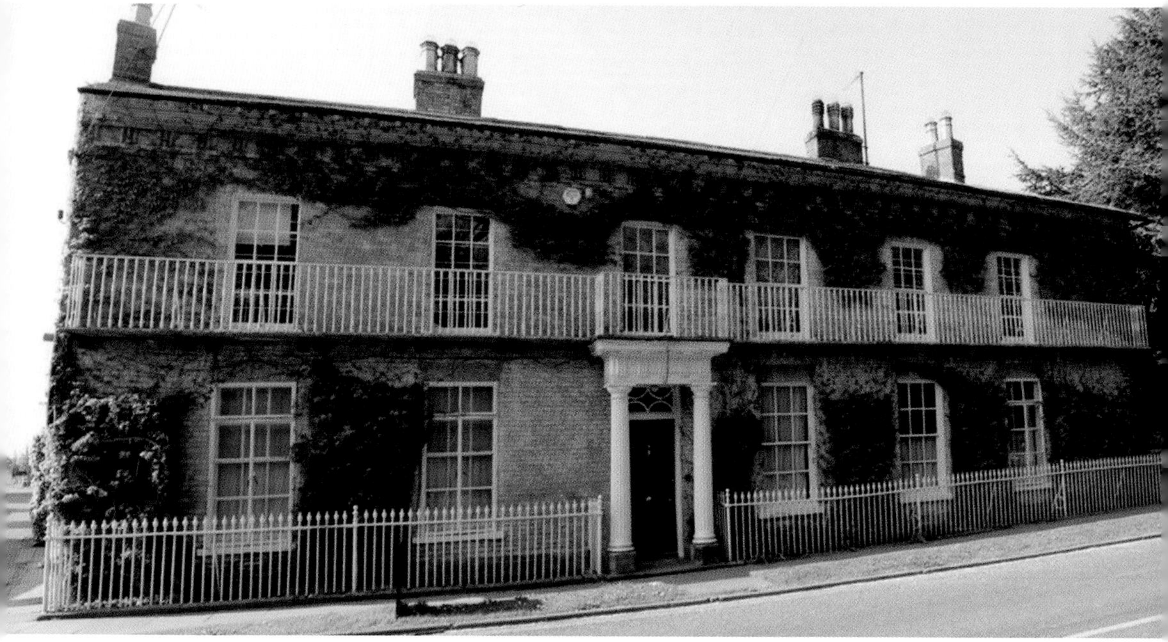

Georgian Magnificence
Salem House in Wainfleet, the graceful and elegant headquarters of Batemans for many years. The building was built in 1825 and is Grade II listed. The wrought-iron columns and classical doorway with its Doric columns set the whole thing off beautifully. (Jo and Steve Turner via Geograph)

traditional story handed down over the years is that Susannah used to brew beer back in the farm kitchen at Friskney for their farmhands at harvest time. If so, she would have been following a centuries old tradition of home brewing, of perhaps variable quality. It's also said that Edwin arranged for his existing brewer, who was blind, to stay on to help George (as my invented discussion suggests). This brewer apparently had an acute sense of smell and taste and was able to check temperatures by dipping his elbow in the brew (blind people do not, in fact, have heightened senses to make up for blindness, but they may have learned to use their remaining senses more effectively). If he could keep his elbow in the brewing liquor without pain the temperature was too low, and if it was very painful it was too high (subsequent fully sighted brewers sometimes adopted the same method).

There is some circumstantial evidence to support the blind brewer story; Edwin Crow definitely had a blind groom (a newspaper report of 1852 records Crow's groom, 'who is quite blind', finding a lost sovereign at the brewery when he heard it clink against his foot). It is quite likely, in fact, that his name was George Harness, who died at Wainfleet in 1881. Sightless since he was five years old, and known by the rather heartless name of 'Blind George', there would have been time for him to pass his trade on to George Bateman and others. How much brewing George did himself is open to question; he referred to himself as a brewing 'novice' as late as 1905 and it may well be that he employed brewers at least some of the time.

The Crow brewery was almost certainly only intended as a stopgap; within a few years, George had moved all of the brewing plant that could be reused over to Salem House, and installed it in the coach house at the rear of the building. This marks the start of the brewery site as it is today, and further buildings have been added over the years as the business slowly began to grow. From the earliest years Batemans also sold wines and spirits, with the Salem House cellars used for bottling and storing whisky, rum and gin. As a further line of business, George tried malting for at least a few years; his farming contacts made him well placed to obtain good-quality barley. He and Susannah also baked bread to sell with the beer. Their customers in the early years were mostly farmers, who would visit the brewery twice a year on the big fair days in early May and October, to pay their bills. Some of them paid in kind, rather than cash, offering meat and potatoes to settle their accounts. Once that was out of the way, the Batemans would put on a huge feast, a tradition that continued until the 1930s. Former employee Mrs Bennett, interviewed in 1986, remembered this twice-yearly ritual:

> There were so many regular customers and they could have as much beer as they could drink. They used to have a 36-gallon cask in the barn for them to go at. Every pub owner bought his wife with him. The landlord of the Three Tuns at Friskney had a big family, he often bought six of his children for a feed.

Slowly, George began to increase the number of landlords he supplied – having your bills paid only twice a year was scarcely ideal. He found some of this trade by buying or leasing

Regal Support
The Royal Oak in Wainfleet. Supplied by Batemans from the earliest years, Harry Bateman took a gamble when he bought it in 1919. The pub closed in 2016 and has since been converted into holiday cottages. 'Royal Oak' has been a popular pub name since the Battle of Worcester in 1651, when King Charles II concealed himself in an oak tree in Boscobel Wood to enable his escape from Parliamentary forces. The name came to indicate support and affection for the monarch. (Batemans)

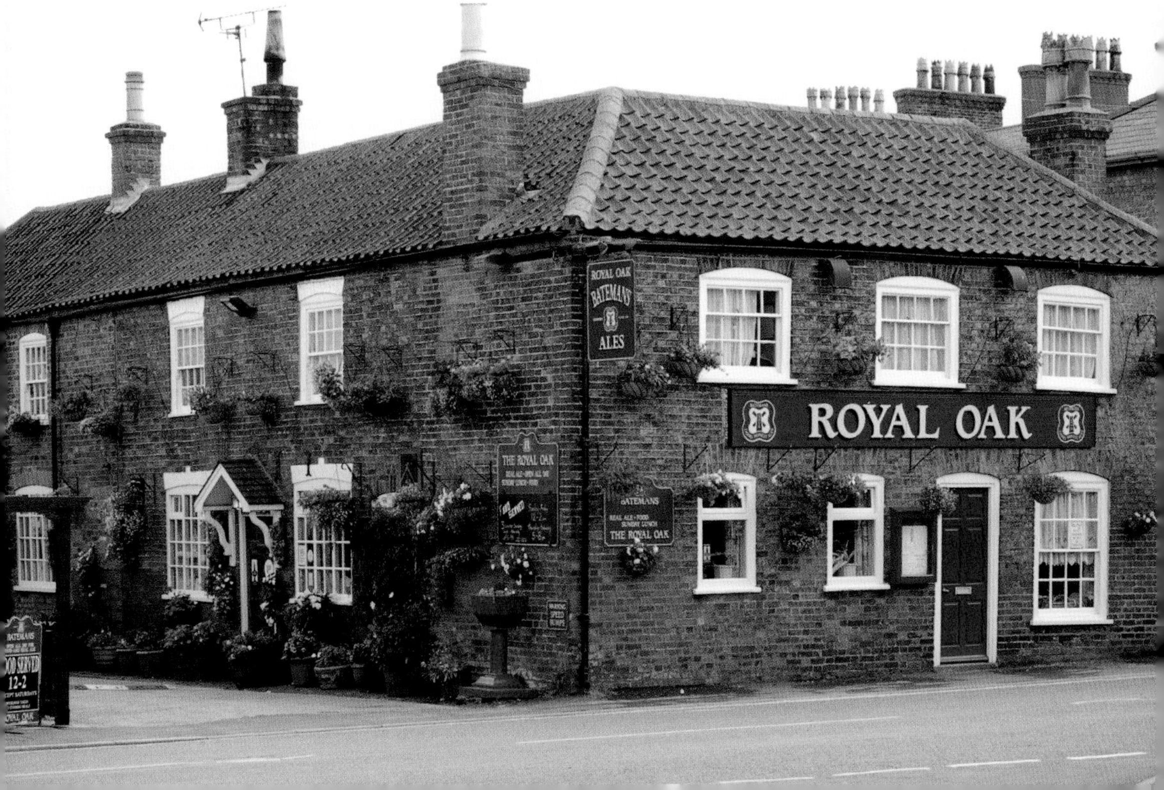

pubs; the Angel Inn and the Royal Oak beerhouse in Wainfleet were supplied by Batemans from 1875, probably under leases. George managed to buy his first pub in 1879: the Cottage Inn, a ramshackle beerhouse a few miles away at Toynton St Peter. It was only a small property, as its name suggests, with a fairly high turnover of tenants, but it was a start. Another eighteen years elapsed before George bought his next pub, the Anchor at Friskney, which he must have known well from his youth; it was an impressive brick-built house dating back to 1850, with a full licence. The next acquisition was a surprise twenty-first birthday present for George's son Harry in 1904: the Old Chequers at Croft, which is the oldest continuously owned Batemans pub. Two years later, Harry bought his second pub, the Red Lion in Wainfleet.

George was known to be protective of his tenants, and the following letter from 1876 to a local newspaper, under the pen name 'Watchful', sounds very much like it was written by him:

> Paltry Case. —Sir, — On the morning of last Sunday week an attempt was made by a policeman, disguised as an Irish harvest man, to prove the sale of intoxicating liquors at the Chequers Inn, Croft, but, notwithstanding the mean dodge performed, the case entirely broke down, as will be seen in your Spilsby Petty Sessions report. Such tricks are very annoying to the proprietors of public-houses ...

The business ended the nineteenth century supplying twelve houses, of which ten were leased from other brewers or landowners.

Long-lived
The Old Chequers, on the banks of the River Lymn at Croft village, near Skegness, has been owned by Batemans for 120 years, longer than any other of their houses. (Batemans)

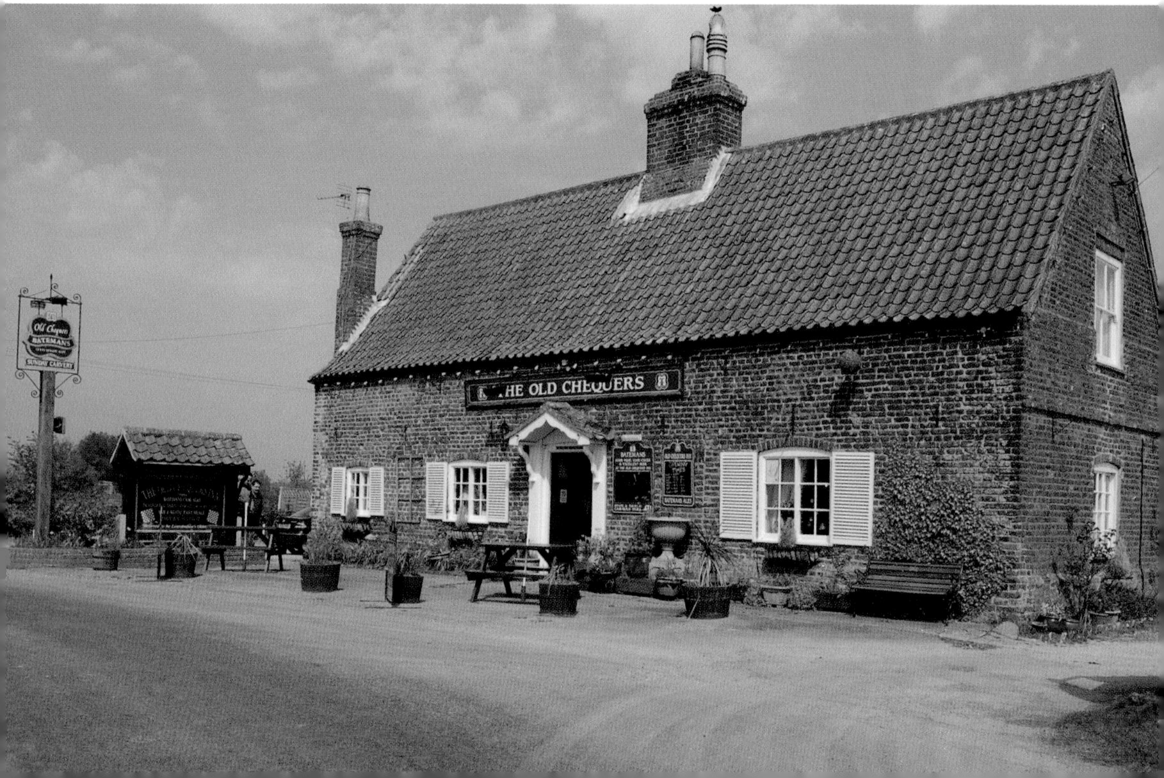

3

The Rivals of George Bateman

Throughout the eighteenth and most of the nineteenth century, brewing in the United Kingdom was a largely small-scale business, but one carried on by many pub landlords and brewers. It was only as the Victorian age progressed that much larger commercial brewers started to appear in the larger towns and cities. In 1874 there were twenty-four other breweries within a 20-mile radius of Wainfleet, as well as a myriad of home-brew pubs meeting their own demand scattered across the countryside. Over the next sixty years all but one of them would close, leaving only Batemans and their much larger competitor, Soulbys of Alford, as the only brewers in east Lincolnshire until Soulbys themselves closed in 1951. What was it about those other brewers that rendered them unable to stay the course? And what was it about Batemans that enabled them to ride out those storms? The answers are complex, individual and varied, and some study of the brewers themselves will help to answer that question.

We'll look first at the Hortons of Old Leake, 8 miles south of Wainfleet, whose dynastic wrangling, linked to water supply issues, eventually proved a toxic cocktail. The patriarch, George Goe Horton, was landlord of the Bricklayers Arms just outside the village on the main road (now the A52). He started brewing in an outbuilding there in the late 1840s, like many home-brew pubs of the time. He'd hoped to pass the business on to his two sons, George and Samuel (George was the oldest by a year), but they fell out spectacularly in 1857 culminating in one of them – perhaps both – throwing a chair at the other in the Ram Inn, Boston. Samuel took himself off to Wrangle, where he opened the new Wrangle Steam Brewery, and George remained at the Old Leake brewery, which he expanded.

Problems with water quality became apparent in 1881, when Samuel was charged with adding salt to his beer. 'Adulteration', or 'bobbing', was something of an obsession with the late Victorians; whilst it was fairly widespread, it was a largely harmless practice and most common in London. It was true that by watering down the beer, or adding salt or sugar, or all of those things, brewers and publicans could make it go further and gain more profit; as a helpful side effect, salt would also make drinkers thirstier. In this case the local police bought some beer directly from Samuel Horton's wife at the Wrangle brewery. The case was dismissed, apparently on the grounds that the local water contained high levels of salt in the first place, but it must have caused some damage to the brewery's reputation. After

Samuel's death in 1882 the Wrangle brewery passed through a number of owners until it was finally closed.

Back at Old Leake, George Horton soldiered on into his eighties, but must have been keen to retire as he offered a substantial loan to George Fordham of the nearby Firsby Brewery Company, to help him buy the business. Fordham soon defaulted on the loan – the Firsby company itself ceased trading a few years later – and Horton instead sold out to Soulbys of Alford, who were only interested in Horton's few pubs and immediately closed the brewery. It was clearly in a poor state as materials from it were being sold off a few years later.

In addition to family squabbles, defects in leadership and oversight often combined to cause problems. Towards the end of the nineteenth century the emerging large brewers, with nationally recognised brands like Bass and Guinness, were becoming more willing to take legal action to defend themselves against brewers who tried to pass off their own brews as better-known and probably superior ales. This was common knowledge amongst the brewing community, but it did not put off Matthias Wharram of the Kiln House Brewery of Burgh le Marsh, who was caught twice in nine years passing off his own beers as first Bass and then Guinness. The first time round, in 1886, he blamed incompetent staff and was fined, and the second time he got away with offering a grovelling apology and paying court costs. Wharram was by no means unique: Bass also threatened Harry Bateman with legal action over his use of a red rectangle on bottle labels for BBB strong ale, claiming it was too similar to the red triangle device they had trademarked. Harry settled privately with Bass, but Wharram's failure to learn from previous errors suggests he was very probably acting deliberately.

Very few of these brewers had any training or experience; most were farmers or publicans who branched out into brewing. When problems cropped up they had little or no prior knowledge to help them. Even George Bateman's brother Francis tried his hand at brewing, a prime example of an ill-judged entry into the industry. After their father's death, Francis had spent more than a decade in what is now Guyana, where he married and had a family. He returned to Lincolnshire in 1888 and rented a brewery in Gosberton, a large village some 30 miles south-west of Wainfleet. Whether George suggested the idea to his older sibling – who was two years his senior – or Francis just decided that he could do as well out of brewing as his brother, we don't know, but the whole venture was a complete disaster. It proved impossible to sell or drink the beer he brewed, nor did he seem to have much understanding of the tied house system, all of which suggests that he had not sought his brother's help. Francis Bateman retreated to Wainfleet, where he became first a butcher and then the town postmaster, though neither George nor Harry seem to have had much to do with him.

George had competition in Wainfleet itself. The Gunson family had been brewers and wine merchants in the town since at least the 1850s, operating from The Vinery, an impressive property on the High Street. John Gunson was the first licensee of the refreshment rooms at Skegness railway station for a brief period, but in 1878 he and his father Isaac bought the Woolpack hotel in Wainfleet marketplace. Isaac and John died within a couple of years of each other – John was only forty-six – leaving the business in 1887 to the twenty-one-year-old William Gunson, who had to abandon his plans to become a vicar. William ran the business until he closed it down on his retirement in 1936, but it's unlikely that he took much business from the Batemans. Relations between the two

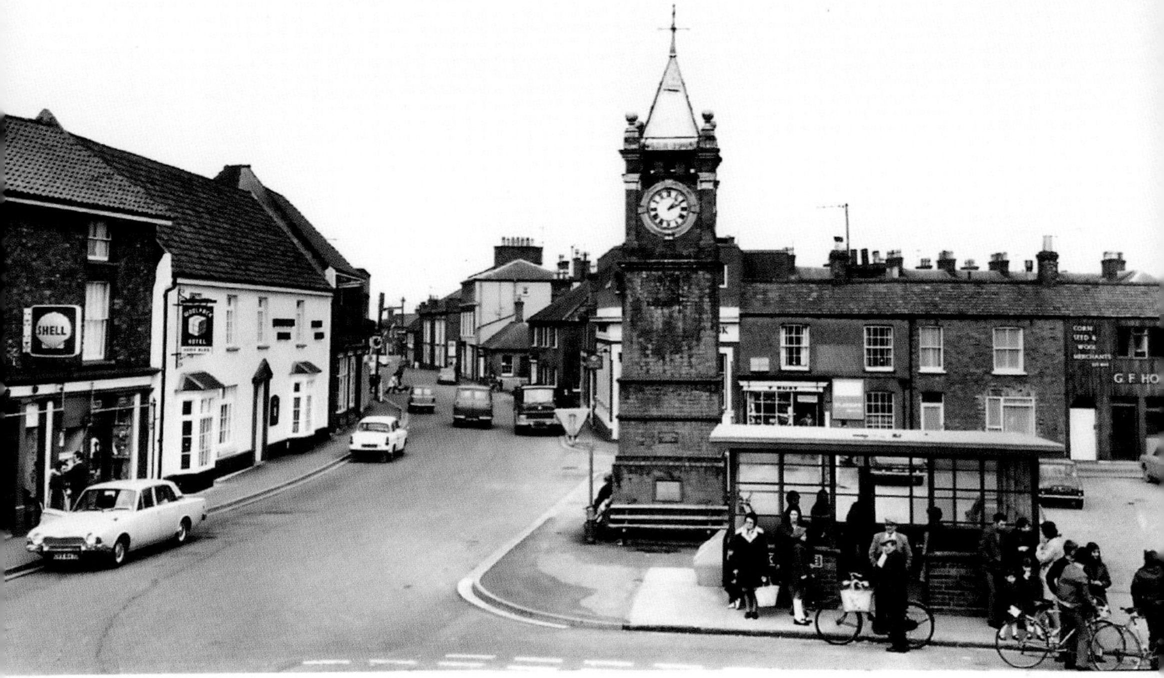

CLOCK TOWER, WAINFLEET

Waiting for the Bus
Wainfleet marketplace on a 1960s postcard, with a number of passengers at the bus shelter wondering where their Lincolnshire Road Car service to Skegness has got to. The town had more shops back then, but Storr's shop remains open, still selling petrol from a single pump set into the shopfront. To its right is the Woolpack pub, then owned by Home Ales of Nottingham; Batemans finally took it over in 1989.

were certainly amicable in 1904, when the Gunsons' brewer died; George Bateman went out of his way to offer help with brewing and bottling, and later Harry Bateman was on first-name terms with 'Will', who by 1917 was already showing signs of wanting to retire. William's main interests may well have remained with the church – he was organist at the town church for thirty-three years and helped to fund the construction of its bell tower.

Another sorry tale unravelled in the market town of Horncastle, at the foot of the Wolds to the west of Wainfleet, where the Phoenix Brewery staggered through at least six changes of ownership in some seventy years, at the end of which Batemans mopped up the few remaining pubs. The founder, Robert Clifton Armstrong, was originally a timber and wine merchant who moved into brewing around 1856, employing brewers rather than undertaking any brewing himself. By his death in 1891, he had grown the business to eleven pubs and a 15-quarter malthouse. The business was sold at auction to a Theophilus Levett, who actually had some brewing experience – he was related to the Bass family and had trained at Burton-on-Trent. He brought a short period of stability until the brewery passed in 1901 to Charles Howland, aged just twenty, who became embroiled in a ham-fisted attempt by another local brewer, Harold Trinder, one of the directors of the nearby Spilsby Brewery, to merge the two breweries. The plan was that Howland would subscribe and pay for shares in the Spilsby company, but Trinder planned to use Howland's money to buy out his fellow directors. Fortunately for Howland, Trinder was a terrible businessman. Blundering spectacularly by paying over £1,000 for a pub in Boston that the licensing authority closed two years later was just one of his lowlights, and unsurprisingly Trinder became personally bankrupt shortly afterwards. Howland had had enough and put the brewery and its thirteen houses up for sale early in 1908.

Howland's successors fared little better. The next owner, Ernest Craven Sharpe, the son of a local wine and coal merchant, employed John Shaw Lord as his brewery manager, a man who had jumped ship from John Marston's Sleaford brewery. Sharpe was insolvent by 1912, and the business was auctioned off the following year to William Baker, owner of Horncastle's Bull Hotel. Baker had grand plans: in 1916, with the support of capital injected by local farmers, he acquired the California Brewery, 5 miles away at Kirkstead, formerly in the hands of the Atkinson family, who for many years had been owners of the Railway Hotel there. Both breweries were hived into a new company, the Horncastle and Kirkstead Brewery Company Ltd, but brewing ceased in the early 1930s. By this stage the tied estate had dwindled to three pubs, which were auctioned in 1933, Harry Bateman being the successful bidder. Despite the moribund state of the business, two of the three pubs are still trading to this day, suggesting that the pubs were not the problem.

What can we learn from these accounts? First, most of the ownership changes were unplanned, involving auctions rather than orderly transfers behind the scenes. These public displays of business problems led to brand damage and staff uncertainty, and because the local press published sale prices whenever possible, the value of the brewery and its pubs was plain for all to see. Second, many of these businesses were quite badly run, the owners making poor business decisions and having a tendency to sail close to the wind, Wharram being a prime example. And family disagreements over policy and direction played their part – as we will see, Batemans were not immune from that. On the other hand, some of these problems were sheer bad luck and could not have been reasonably foreseen. George Bateman did have some good fortune, but he was also a shrewd and quite cautious businessman, qualities that helped him to avoid the pitfalls of his local competitors, quietly flourishing whilst they foundered. Finally, the concentration of brewing in Lincolnshire in the hands of fewer firms was a reflection of national trends; the number of breweries in the UK had been declining since the 1870s, whilst overall output continued to grow slowly. At the turn of the twentieth century, a number of larger firms continued to grow and assert their authority, providing a different sort of challenge to smaller brewers like Batemans.

4

A Blossoming Business, 1900–18

It was George's son Harry who took up the reins in greater earnest, building the brewery up over nearly seventy years to a successful business serving over 100 pubs. Harry was born at Wainfleet on 16 July 1883 and educated in the town at the historic and highly impressive Magdalen College School. Very promising students were given 'exhibitions' (essentially, financial assistance) to its namesake college in Oxford. Harry was not among those chosen, although as a recipient of several class prizes, he clearly was a diligent student. Away from school, he qualified in shorthand. He was showing some entrepreneurial skills by the age of sixteen, selling eggs at the brewery ('from splendid birds on unlimited grass run', proclaimed his advertisements in the local paper). He started helping his father's business in 1904 by taking on some of the office administration and pub estate matters, such as getting on the road to visit tenants on George's behalf, and ordering new pub signs – 'be

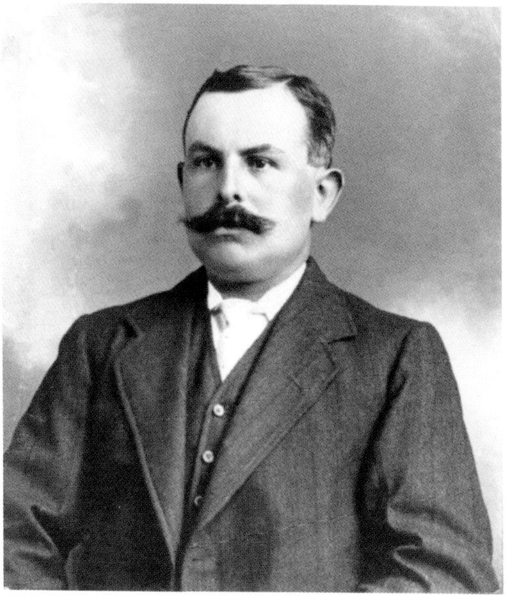

Driving Force
Harry Bateman was the man who took the brewery from a prosperous but small business to one with over a hundred pubs across a wide area. In this formal portrait he appears to be in his late twenties or early thirties, which would date it to around the start of the First World War. He wears a three-piece suit and has a full moustache, possibly as a patriotic gesture in solidarity with servicemen: in the British Army, moustaches were compulsory until 1916. (Batemans)

as economical as possible', Harry warned in one letter. We are very fortunate that many of the letters sent by George Bateman and his son Harry have survived, as have some of the replies. Though now fading and fragile, these documents can still reveal a rich and insightful picture from over a century ago of the fortunes of a small country brewery and its owners – and ultimately how the course of the First World War caught them up in a whirlpool of uncertainty, which seemed at times to threaten the business's very existence.

Harry became a partner in the business in April 1907, but it was not inevitable that he would take over the brewery. Although he continued to help out as needed, in 1905 George had given him a bakery business, house and off-licence near to the station. It seems quite likely this operated from the former Crow brewery site down by the river. Harry ran the bakery for at least three years and possibly longer, but found that the brewery took up more and more of his time, especially after his father suffered a leg injury. He advertised for tenants on several occasions and seems eventually to have closed the business in 1916. Harry was also involved with the local cattle market and farmed pigs for a while – he was busy even then!

For a business as small as theirs, finding reliable new tenants when the old landlord left was crucial, and the Batemans were firm believers in due diligence. Much of the time their approach worked well, but some people caused endless problems. When Elizabeth Hudson, the long-serving landlady of the Anchor at Friskney, retired in 1904, George was anxious to find 'good, respectable people' as her replacement. His choice from several applications was a local man named Bartholomew Simpson, whose father and others had provided good references. Simpson and his wife Magdalene had not only been running the Royal Tar beerhouse in Brentford, Middlesex, but also a local dairy. However, within three months of moving in, George had to tell him, in a stiff letter, that there had been complaints of drunkenness at his house, and if he did not keep better control of his customers ('the language used is a disgrace') he would be removed. George well knew that licensing magistrates could seize on alleged drunkenness as a pretext for closing pubs, and strongly suspected that the vicar of Friskney was behind the complaints. Despite these early problems, Simpson lasted ten years at the Anchor, his parting shot being to complain that he couldn't make a living there. He started a sideline as a poultry dealer, and Harry unwisely lent him £50; when repayment was due, Simpson told him: 'Do what the ___ you like, you can't get it.' Simpson promptly bought himself a new car and was not surprisingly declared bankrupt shortly afterwards.

Other tenants were difficult too. One of Harry's earliest business encounters was a spat with Alice Spence, the landlady of the Shepherd's Arms beerhouse just north of Wrangle, a roadside pub the Batemans had bought in 1904. Harry found her disrespectful: 'I have had more abuse from you during the last 18 months than from all our other customers and tenants', he told her in May 1906, refusing to work with her in future and suggesting she should look for a new house. Like a number of Harry's threats, he seems soon to have cooled down and the Spences remained at the house for several years, Harry even being prepared to ask the local police whether there was any prospect of Spence's son Samuel obtaining a pub licence for the Shepherd's Arms.

Both George and Harry would help their tenants if they possibly could. Alf Shearman, an old family friend, was also landlord of the Red Lion in Wainfleet, but around 1915 was clearly suffering some problems. Harry found that speaking with him had no effect, so he wrote a dramatic letter saying that the situation 'cuts me like a knife' but Shearman faced 'a Drunkard's Death' and eviction if he did not stop drinking. His words may well have helped, as Shearman held the Red Lion for another thirty years and served with Harry on a local publicans' committee. The Batemans were equally prepared to help brewery staff; when Charles Hanson left in 1910, George sent him a letter addressed 'Dear Charles' – in an age when familiarity was frowned upon in business correspondence – expressing his thanks for his thirty-three years with the firm, and enclosing a cheque 'as a token of my great appreciation'.

When David Dawson of the Cottage Inn at Toynton was having problems with drunken customers in January 1908, he was advised to 'think more and say less'. These issues led to the licence of the house being withdrawn by the magistrates in 1910, to George's regret; he'd owned the pub for thirty years and said it did a 'nice trade', although he was pleased with the compensation received. This payment arose from a system established in 1906 that made it much easier for magistrates to close pubs, mainly in response to campaigning by temperance groups. All local brewers and publicans were required to contribute to a central fund, supposedly to provide a financial cushion for the landlord and pub owner if the magistrates refused to renew a pub's licence, although it was rarely enough: 'closed under compensation' became the bane of brewers' lives and many had their estates badly affected by what was often a target-led campaign by police forces and magistrates. Batemans, with their initially small estate, only ever had four pubs closed under compensation, unlike several of their competitors; after the Cottage Inn, the next compulsory closure was not until 1956.

Father and son were not always consistent. In May 1906 George Thompson, a local farmer, asked Harry for a long-term loan. Although Harry reluctantly declined, he was still prepared to lend him £10 to tide him over for a few months. The Batemans dismissed a clerk called Watson in 1910, only to take him back on condition that he was 'more polite, considerate and obliging to the members of the Firm, the family and the customers, and to be punctual at all times'; he lasted another two years after the reprieve. Maybe they were sometimes led more by their hearts than by their business minds; in short, they were men who believed in helping their local community.

As well as these vignettes from brewery life, we can also detect some rough outlines of what George and Harry were like as people, away from the brewery. As well as his status as the owner of a local business, George already held a number of partly voluntary roles in and around the town, and would have been well known amongst the wider farming community in east Lincolnshire. This naturally made him something of a lodestar for local people, some of whom still could not read or write, and it is clear from the letters he wrote that George, and then Harry, were men who were happy to help others, sometimes in the most difficult of circumstances. In the midst of war, for example, Harry wrote to the army in March 1917 to enquire about Private J. Gilliatt on behalf of his mother, who believed her son was missing. She was right: sadly, he had been killed in action. Six months later Harry was trying to get medical help for the 'helpless condition' of his tenant, Mrs Alice Robinson of the White Lion at Little Steeping. If he couldn't book her a room at Buxton, famed as a health resort, Harry was determined to get her to the more local facilities at Woodhall Spa instead and help financially.

Then there was the sorry tale of the dodgy dentist and Mrs Bee's false teeth. An unregistered dentist called Joseph Donaldson traded briefly from premises at Scunthorpe in 1915 and local widow Mrs Bee ordered a set of dentures from him. When they failed to arrive the toothless woman consulted Harry, who was the executor of her late husband's will. Harry fired off a series of increasingly annoyed letters which eventually prompted the arrival of replacement teeth, although only the lower set fitted properly. The lack of further correspondence suggests Harry's insistence won the day, whilst Donaldson – under a number of assumed names – continued to prey on luckless customers up and down the country, until declared bankrupt.

Relationships with other breweries could be fractious at times. Around the start of the twentieth century George had a running sore with Soulbys of Alford, going so far as to tell one landlord that they were 'not to be relied upon', although he said the friction ended when Soulbys sacked their brewery manager. It was more frustrating to deal with Thomas Salt & Co. Ltd of Burton-on-Trent, a much larger firm whose pub estate sprawled untidily across the Midlands, the result of several mergers. Tucked away on the easternmost fringe of their empire was the Good Intent at Friskney. Nearly 100 miles from Burton and rented to Batemans since the 1890s, it was a case of out of sight, out of mind, and George was struggling to interest Salt's in repairing their increasingly tatty pub. By 1903 several window and door frames and the cellar floor needed replacing, but nothing was done for at least a year, an interesting insight on the state that some pubs were allowed to get into at the time. George told Salt's that 'the place really is a disgrace' and eventually threatened to give up his own tenancy unless there was a rent reduction. That must have provoked some improvements because by 1917 Harry was unsuccessfully pressing Salt's to sell the pub to Batemans for £500 (they eventually bought it at auction in 1928 for £450). Despite all this they sought advice from Salt's from time to time both on brewing and pub licensing, Harry laying on the flattery in 1911 by telling them 'we are such a small way of business in comparison to you'.

George and Harry's relationship with their suppliers was at times quite adversarial, and it's clear they expected good value for their money. They spent some years in verbal jousting with their maltsters, a large firm still trading today. As both a farmer and brewer, George was well placed to run a critical eye over the malt he received, and he was not happy. 'I consider it very strange that you should not be able to supply me with a uniform Malt. Can you do so if not I shall be compelled to seek a change,' he declared in 1905. A few months later he was again dissatisfied: 'I find a very large percentage of flinty and also damaged and mouldy corns. I should like to know how you account for this.' Prices, too, were a matter of dispute, another increase prompting a letter from Wainfleet: 'If this is your way of doing business, it is not ours.' Eventually George took his custom elsewhere with a final sally: 'You have used us very badly.' Others were on the receiving end too: the engineering firm Llewellins & James had supplied George with a new brewing copper in 1896 but he found it 'far from a good one, and I think there is a great doubt about it ever being fine copper'. There was also a falling out with the brewery engineers Pontifex & Son of London in 1906, the main issue being labour costs of the workers they sent to install a new pump: 'I don't quite see why I should pay for the time they are at meals, viz., 1½ hours each day,' George grumbled.

What to make of these acerbic and at times quite assertive letters? Whilst George may come across as difficult to please, it is worth bearing in mind he was dealing with much bigger firms in all cases, ones that he considered had messed him around. He always explained why he was unhappy and was clear about the outcome he needed. He, and Harry too, expected good value for his hard-earned money, and he made sure that he got it – or he would go elsewhere. This approach helped to ensure that complaints from customers were kept to a minimum; George particularly seems often to have worried about beer quality and hence the brewery's reputation, to the extent of buying some fairly outlandish-sounding concoctions, such as a yeast enhancer called 'Phosphated Diastatic Food' ordered from a chemist in 1904, which helped to clear a batch of cloudy beer.

The onset of the First World War initially had little effect on brewers, but as the conflict ground on, the industry found itself more and more restricted – although as we will see, Batemans bucked the general trend by achieving, through unremitting hard slog, significant growth during this period. The first signs of this squeeze on the industry took the form of successive very large increases in beer duty to help pay for the war effort,

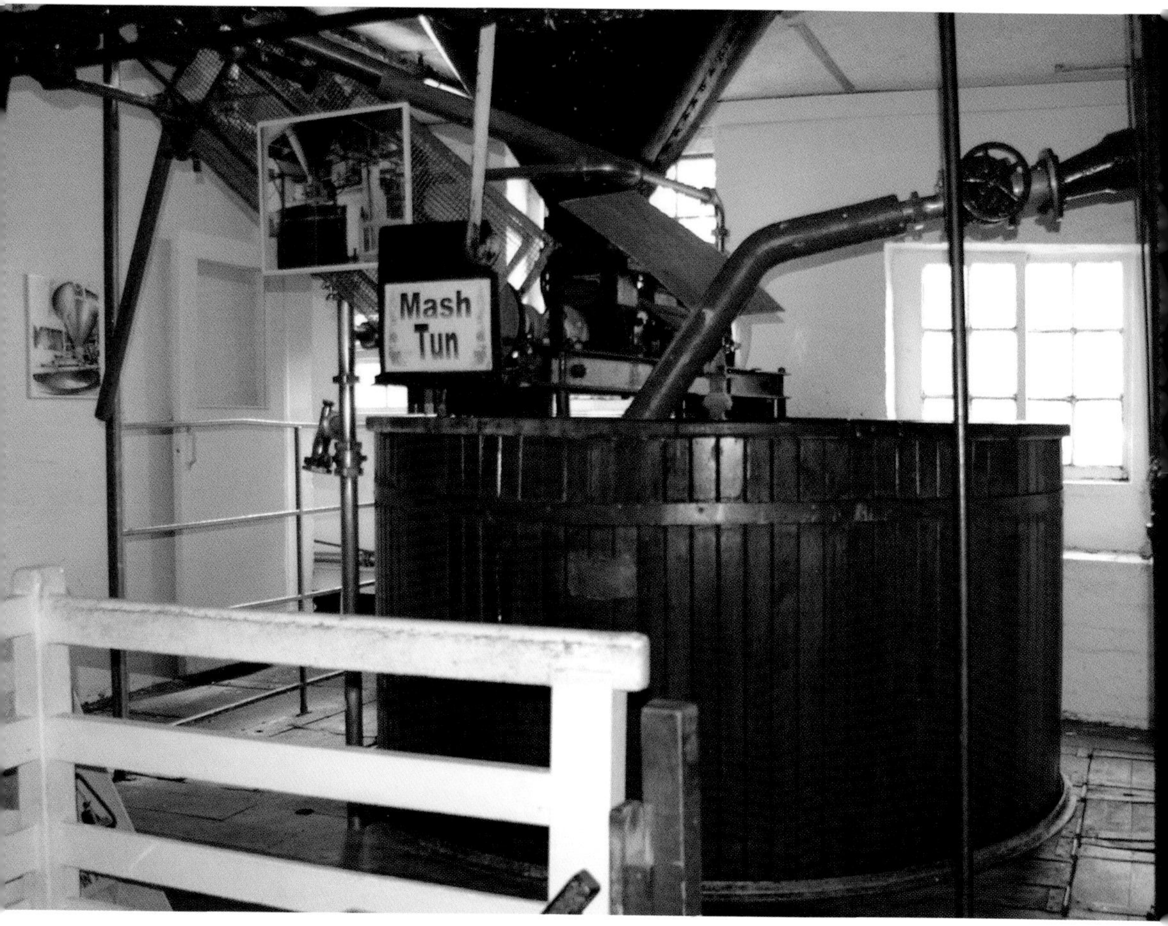

Remarkable Survivor

The mash tun in the older part of the brewery, manufactured by Geo. Adlam & Co. of Bristol in 1901, is still serviceable and is used for occasional brews. The key artefacts in this area are labelled to help explain the brewing process during the very popular organised tours of the brewery. (John Law)

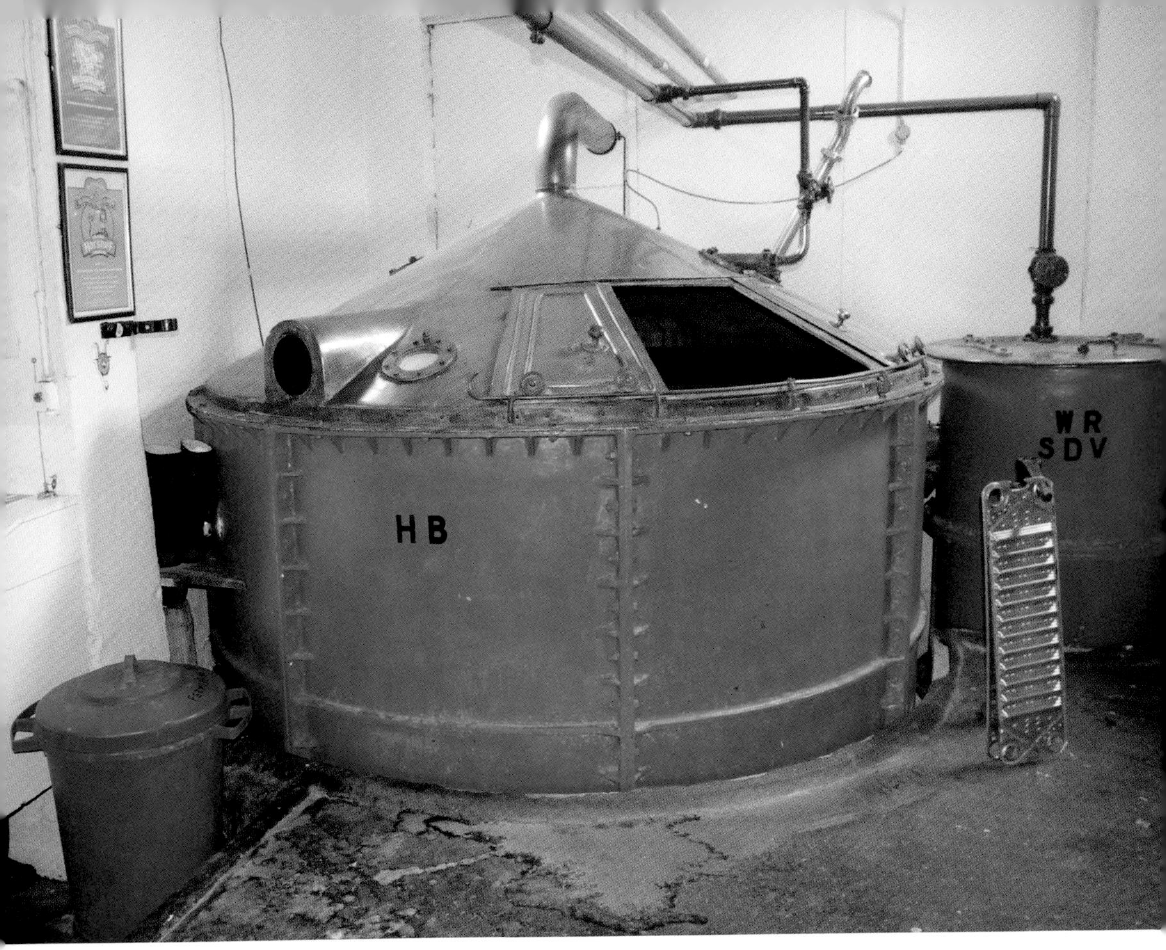

Brewing Blues
The original hop back, with the sugar dissolving vessel to its right. In many breweries the function of each vessel is indicated by initials, as in this case. The blue dustbin is probably a happy coincidence! Photo taken in 2013.

starting in November 1914 when David Lloyd George, the Chancellor of the Exchequer, imposed a threefold increase from *7s 6d* to 23*s* per standard barrel. Harry, a staunch Conservative, had a particular dislike of the Liberal politician and teetotaller, whose attacks on the drink trade were notorious. Opening hours were curtailed, and successive measures had the effect of reducing both the volume and strength of beer brewed. In 1917 the law was further tightened to allow the production of just 28 per cent of pre-war volumes, linked to a certification system that caused significant problems when a brewer sold a pub. Harry spent an inordinate amount of time in testy arguments with other local brewers about their reluctance to provide him with the required certificates, but brewers soon found that they needed to work together if they were to meet the needs of all their customers as the reality of low production levels finally dawned on some of them. Price controls followed in October 1917, with a maximum price of 4*d*. for lower-strength beer and 5*d*. for most stronger beers. For Harry, this meant a greater demand for beer in his pubs and consequently less beer was available for farmers.

The government were prepared to allow brewers to produce beer for special circumstances, which included 'harvest beer' for farmworkers. Wainfleet is surrounded by fertile farmland, so we find Harry in August 1917 applying for a licence to produce fifty standard barrels above the 769 produced in the same three-month period the previous year; interestingly, he was still producing much more beer for his private customers than for pubs. To Harry's annoyance, he was only allowed to brew twenty more barrels, and he complained bitterly to his trade association. Fortunately, he had excellent support from Henry Hobson, the Nottingham-based regional chairman of the National Trade Defence Association. Hobson helped Harry negotiate the challenges of wartime production to the extent that the two struck up a firm friendship, which saw Harry sending food parcels to his friend and his young family to help with wartime rationing. There was also a scheme, from 1917, to permit brewers supplying workers in the munitions industry with lower-strength beer, a tacit admission that earlier policies had not worked, to the extent of causing social unrest. As there were a number of factories given over to the production of aircraft and other military hardware in Lincoln, Scunthorpe and Grimsby, Harry applied in October 1917 for a licence to produce 200 extra barrels of 'munitions beer' (as he called it) – 'we consider this quantity is about half of what is really required' he commented, perhaps tactically. He was only granted a licence for thirty-six barrels, although that rose to 180 in the next quarter, once again with advice and assistance from Hobson.

But there were worse things to worry about. The industry collectively was deeply concerned about the threat of nationalisation. Already a vast swathe of pubs, totalling 363, across the Carlisle and Gretna area in Cumbria and the southern tip of Scotland had been compulsorily acquired by the government in June 1916 as what became known as the State Management Scheme, followed by pubs in Enfield, London, and Invergordon, Scotland. This had been largely motivated by concerns about drunkenness at the vast munitions factory in Gretna, where huge quantities of explosives were handled, but there was the lingering suspicion amongst brewers that here was more evidence of the power of the anti-drink lobby, with Lloyd George a malign, controlling hand now that he had been moved to Minister of Munitions. This clearly caused Harry great anxiety, prompting him to write to the managing director of the Maryport Brewery, near Carlisle, asking them how their houses had been valued by the government 'as we expect eventually we should also be bought out'. Those fears were not realised. The wartime economy could not have supported more nationalisation, and the government came to realise the drink trade could be better managed by measures such as price controls. But to Harry, struggling with the uncertainty of wartime, the threat must have seemed very real. His father George had never fully recovered from a serious accident to his leg sustained when alighting from a train at the local station back in 1908, so that he could take little active part in the business, although he continued to offer advice and support where possible. In February 1918 Harry bought his father's share of the partnership for £4,299, explaining to his solicitor that 'my Father desires to withdraw from the business'. The word 'retirement' was not used, and it is quite likely that George, then aged seventy-four, remained available in an advisory role. The price included the brewery itself and two pubs that were still in George's name (the Anchor in Friskney and the Chequers at Croft).

Harry was also trying to cope with a gradually reducing workforce as the demands of war took their toll. Until January 1916 service in the armed forces had been voluntary, but with the increasing toll of lives, conscription had to be introduced. There was a system of exemptions, principally for people in work of national importance or those for whom call-up would cause significant business or domestic hardship. Did brewing fall under either of those categories? Harry hoped that it did, both for the sake of his own workers, and for him. At thirty-two years old, he was well under the age limit of forty-one for call-up (which had been extended to married men in May that year), although he did serve as a 'parish constable' for Wainfleet, effectively a reserve policeman in case of invasion. Because many of Batemans' workers were called up, as the months progressed the workforce gradually became more and more depleted. For instance, the clerk and cashier for the last four years, Stanley Robinson, aged twenty-two and 'almost indispensable to us', was called up in September 1916 despite having been rejected by a military doctor eight months earlier on the grounds of flat feet and a broken toe. Harry nevertheless tried to help Robinson by getting him into the fledgling air force. Harry faced the prospect of having to do all the office work himself, but luck was on his side: a young woman just out of school had found out about Robinson's departure. Over fifty years later, Mrs Bennett recalled:

> I was top of the form, brave and eager to do something, so I called into the office to see Mr Bateman. I started the next Monday morning. I borrowed 50 shillings from my father to buy my first bicycle and cycled to the office every morning. Of course, it was all new to me but Mr Bateman was very good, he showed me all the books, every barrel had a number and had to be taken down. He gave me some letters to write, I scribbled them down in longhand and them wrote them ready for him to sign. We got on famously and eventually I looked after all the money when the draymen came in. I was in the first-floor office leading into the hall of Salem House, we had a bell in the corner and if Mr Bateman was in and someone wanted him I just used to ring the bell. There were only the four of five of us on the premises.

As the war dragged on, so the military tribunals which presided over exemption claims became less inclined to allow them. Although his brewer Ernest Riggall had been given a conditional exemption from service, in April 1918 that was rescinded and he was called up the following month at the age of thirty. By September, Harry was leading brewing operations himself, with the support of draymen John Smith (forty-five) and Charles Wright (forty-four), who had both been working much of their time in the brewery, with 'the parish sweep and an old road mender in their spare time'. Harry's sister Jessie, a few years younger than him, also helped in the office. As Harry had feared, his own exemption, agreed in February 1916, was then reviewed. He put up a strong argument that he was a sole trader, supplying forty retailers and brewing ninety barrels a week. Harry attended a military tribunal in Spilsby at the end of May 1918, clearly very nervous of the outcome, and launched into an explanation of his position, but was cut off after a few minutes. To his great relief, his exemption was again confirmed. Wider problems such as depleted staff, price and volume restrictions affected all brewers and there was some degree of co-operation; for instance, a local brewer (such as Batemans) might agree to supply the houses of a more distant owner.

Harry spotted an opportunity itself in the pubs and especially working men's clubs in Lincoln, Scunthorpe and Grimsby, where membership had increased rapidly before the war – some clubs were approaching 1,000 members. Whilst club licensing was not as strict as pubs, wartime restrictions meant that clubs were trying desperately to get hold of even a few barrels of beer to meet demand. The location of Batemans' brewery, so close

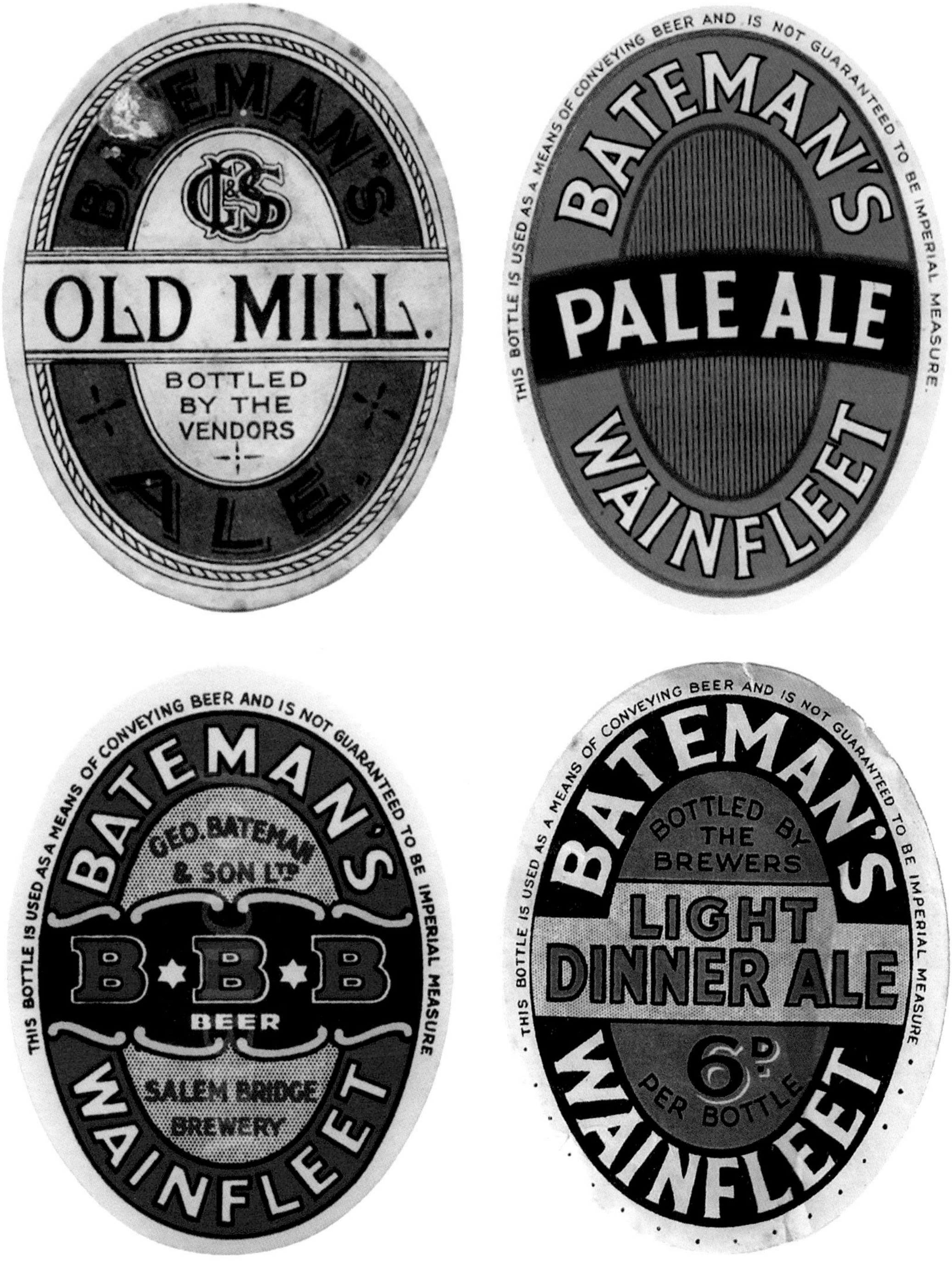

Early Twentieth-century Labels

A selection of beer labels that show the variety of bottled beers Batemans produced. Old Mill (top left) was first brewed shortly after Harry Batemans bought the Salem Bridge windmill in 1920 and was discontinued in the 1940s. Pale Ale (top right) was an earlier but longer-lasting brew. BBB (bottom left) was Batemans' strong barley wine, and Light Dinner Ale (bottom right) was a lower-strength beer. In all four cases the designs are bright and eye-catching despite the limited use of colour.

to Wainfleet railway station, meant that it was relatively straightforward to send bottles and barrels of beer some distance by rail. The challenges were being able to brew in the first place, and hoping that the railway company – depleted by wartime staff shortages – didn't lose the consignment. By September 1917, Harry was brewing for the Lincoln Arms pub in Grimsby and an off-licence in Cleethorpes, and soon after was supplying a number of army camps in Skegness and at Belton Park, near Grantham. Demand was so great that Harry was juggling customers, especially in the north of the county. We know that in June 1919 Batemans were supplying 420 barrels every three months to eight working men's clubs in and around Scunthorpe alone, to which could be added clubs and pubs in Immingham, Grimsby and Lincoln. He came away from his first visit, in February 1918, to one of the biggest Scunthorpe clubs (the Brumby & Frodingham Working Men's Club) highly impressed: 'This is the place for beer shifting, I never saw anything to equal it.' This was all happening whilst Harry was also trying to supply his existing houses and private customers, brew at least twice a week and write up his accounting books. It was only at the insistence of the local income tax commissioners in 1916 that he engaged an accountant. Until then, as he admitted, 'I had no idea of the net profit of the Firm'. But by the end of the war it became clear that the unremitting hard slog had all been worth it. Batemans was highly profitable, and there was enough money available to finance the expansion of the pub estate.

5

Steady Expansion, 1918–45

If the stresses and strains of wartime had not been enough, Harry had given himself a new challenge. He was keen to increase the size of the pub estate, which in 1916 only numbered eleven houses, mostly low-barrelage pubs within a 5-mile radius of the brewery. Harry's efforts to buy more expensive and prominent pubs had so far been unsuccessful – he had even considered buying the Old Bridge Brewery and its forty-two pubs, 100 miles away in

Time to Wine Down

Ridlingtons, the Boston wine merchants bought by Batemans, were based at this impressive Grade II listed mid-seventeenth-century town house, which Harry Bateman renamed Pilgrim House. Its extensive cellars were ideal for a wine and spirits trade. The ground-floor shop frontage was an early twentieth-century addition, probably by Batemans. The building is now divided into flats. To the right is a passageway leading to the Ship Tavern. (Richard Croft via Geograph)

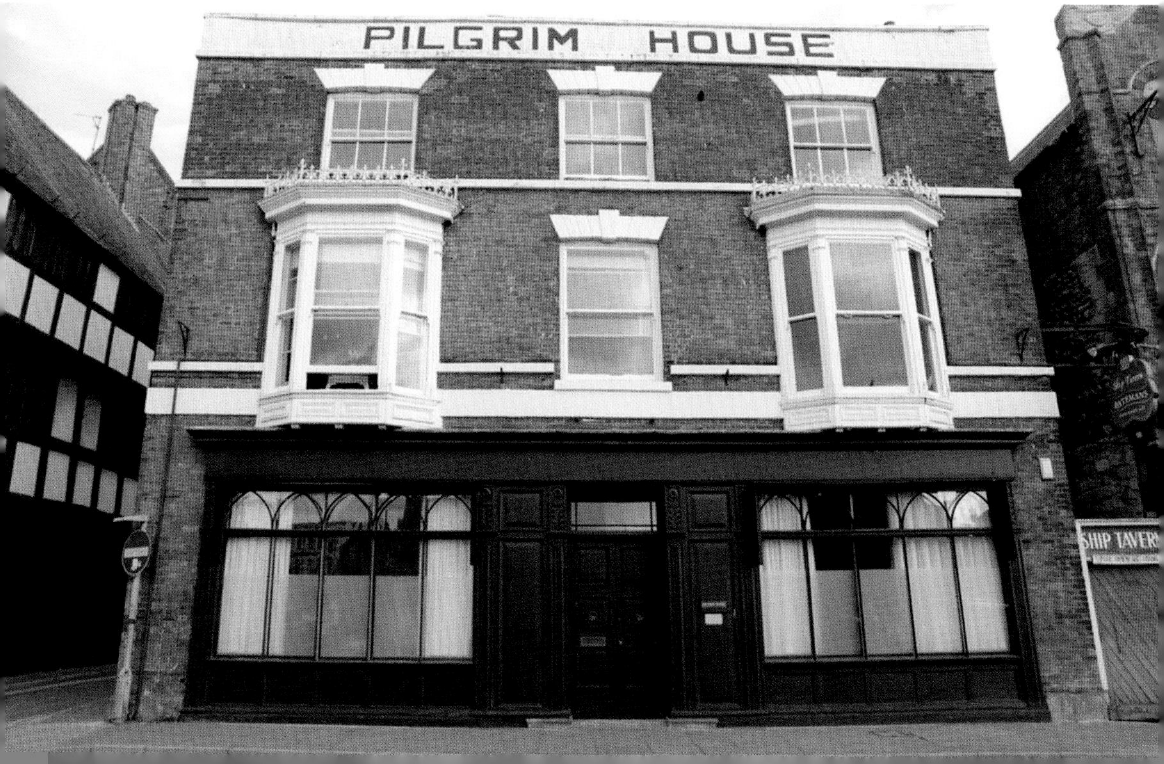

Wakefield – until he heard from a local solicitor that J. E. Ridlington & Son, a well-regarded wine merchants' business 18 miles away in Boston, might be available. The firm operated from an impressive seventeenth-century house on South Street, and also had a bonded warehouse close by on the river quayside and a bottling store as well as a busy pub, the Ship Tavern. The prospect was very attractive. Boston was the nearest major regional centre to Wainfleet, with two large markets a week attracting farmers and businessmen from a wide area across east and south Lincolnshire, and Harry tended to visit Boston frequently for all the latest trade gossip and news. Not only that, but it was a private sale, with no risk of being overbid by larger firms, as had happened on several other occasions. Moreover, Harry believed Ridlingtons – which supposedly dated back to 1705 – had been stuck in a rut for years. Thomas Ridlington, the owner, was sixty-seven and wanted to retire. The business badly needed a shot in the arm, and Harry felt he was the right person to deliver that injection.

Harry spent £2,800 to acquire Ridlingtons in July 1918, with the help of a £2,000 loan from its owner. It was a bargain, and he bought four other pubs that year as the drink trade started to emerge from the war years. He lost little time in putting his own stamp on Ridlingtons, taking on the existing staff whilst keeping his new business separate from Batemans. However, he was plainly unhappy with the performance of the long-serving local manager, John Chambers – 'he is not the man for the position,' he grumbled. Harry found he had to keep an almost daily eye on Chambers and tell him what to do, which cannot have been easy at a distance. After six months Harry let Chambers go, replacing him with a new and younger man, but tragically Chambers took his own life a few weeks later. Considering that Chambers's father also worked for Ridlingtons, and also killed himself, it's strange that nobody seems to have warned Harry, whose correspondence suggests shock and surprise at the news.

Harry's confidence and willingness to take calculated risks was shown in 1919 with the episode of the Royal Oak at Wainfleet. This presentable and profitable town centre beerhouse was rented from a local landowner, and had been one of the first pubs George Bateman supplied. The landlord, George Overton, who had been in charge ever since he left the navy fifteen years earlier, had an unblemished record. Nevertheless, it had been earmarked by the local constabulary in a rather ham-fisted way as the latest target of their continuing campaign to reduce the number of licensed houses in the area. Even though this cloud was hanging over the pub, Harry risked £850 on buying it even before the magistrates heard his appeal to keep the licence. Harry must have been confident of his case, but just to make sure, he got up a petition to keep the beerhouse open.

As the prospect of an end to the war became more likely towards the end of 1918, Harry felt able to advertise for staff, in competition with other breweries trying to rebuild ('if possible send photo in Civilian clothes' he asked one applicant). For the key role of manager and head brewer he decided to take on Cranmer Tom Parkes of Alsager, Cheshire, who came strongly recommended; Harry seems to have negotiated his early release from the Army Service Corps, where he had served as a private. He also considered taking back Ernie Riggall, his former head brewer, but seems to have decided against, possibly due to an incident that Mrs Bennett remembered: 'Billy Dauber, an old man working in the brewing room, let the cat out of the bag about Ernie Riggall. He had been taking the bung out of the barrel and was sucking up it neat with a straw.' By 1921

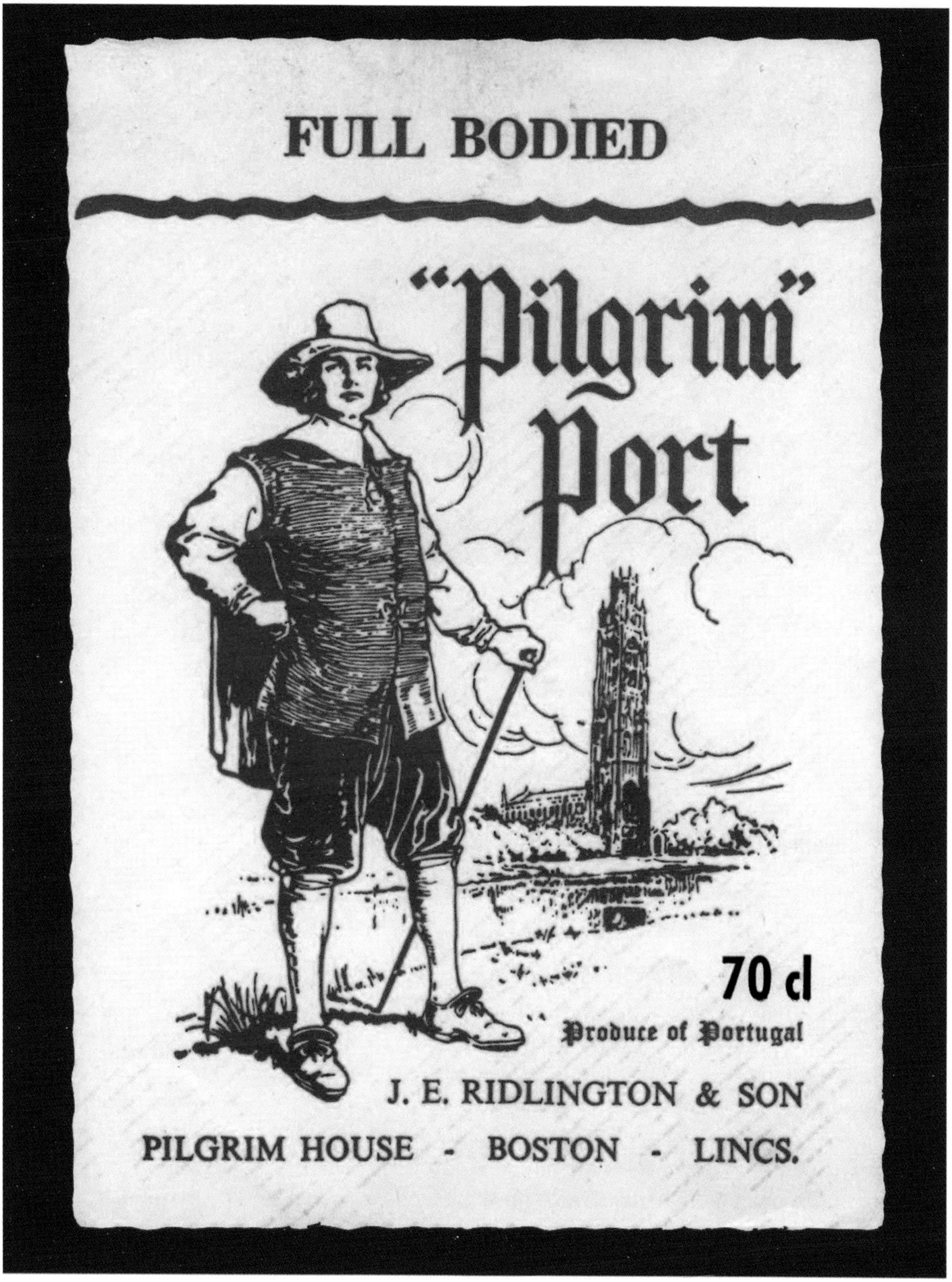

Pilgrim's Progress

Under Harry Bateman's ownership, Ridlingtons adopted this figure of a 'Puritan gentleman' for its labelling. Dressed in period clothing and wearing a 'capotain' hat, he's linked to Boston in this image by the inclusion of the Stump, the town's huge parish church, but its distance from the figure hints at the journey the Pilgrim Fathers made to New England in 1620. (Batemans)

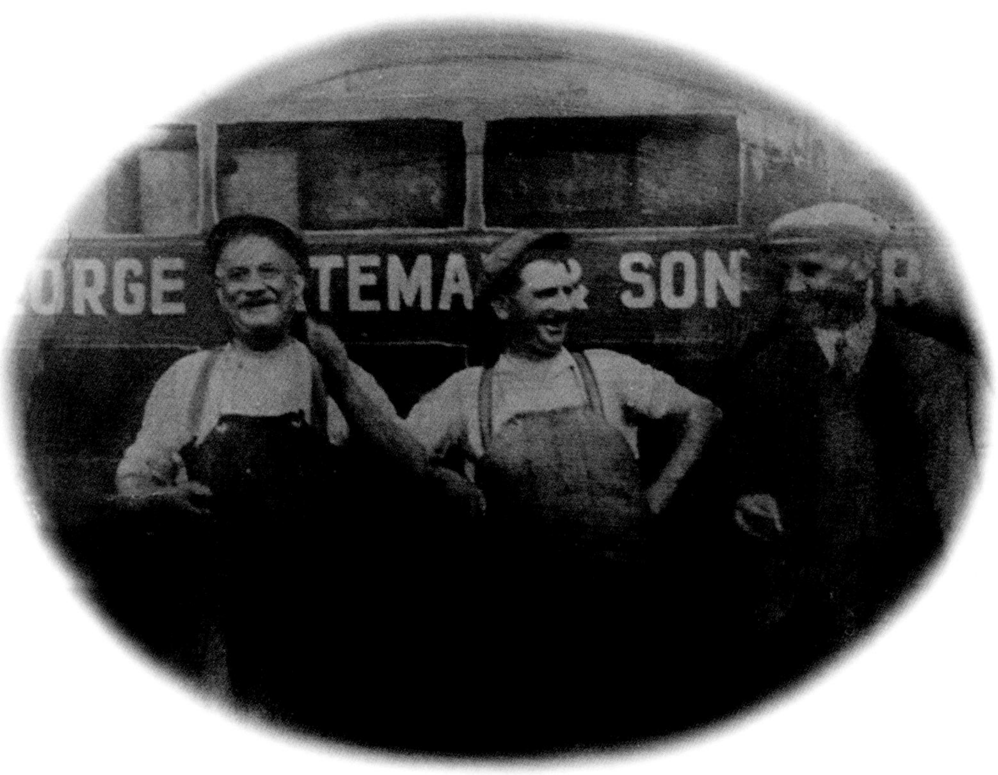

The Three Jolly Draymen
A lovely image of some of the staff from the interwar years. Unfortunately, we don't know who any of them are, but they are clearly having a great time. (Batemans)

Harry had fifteen employees, slightly more than before the war, although only three of them had previously worked for Batemans. The long-serving John Smith was to retire in 1936 after forty-three years with the firm, and William Bromfield called it a day in 1940, but it is interesting to note from that year's census how young many of the staff were: the average age was twenty-nine, reflecting to some extent the hard physical work involved in breweries. Most of the staff were local men, based at the brewery most of the time, but with two draymen and a salesman (known in the trade as a 'traveller') out on the road.

As well as increasing the workforce, Harry also wanted to keep the brewery as efficient as he could afford. Deliveries had always been made by the drayman, latterly John Smith, using a horse and cart. He was out every day (there were no half days!) and often did not return until around nine o'clock in the evening. With motorised transport becoming more reliable and affordable, Harry recognised the scope for timesaving, so around 1920 he bought two lorries, one for Wainfleet and another for Ridlingtons at Boston; the brewery horses were gradually sold. He had already treated himself to a motor car back in 1918, enabling him to reach his customers, especially the more distant clubs, much more easily.

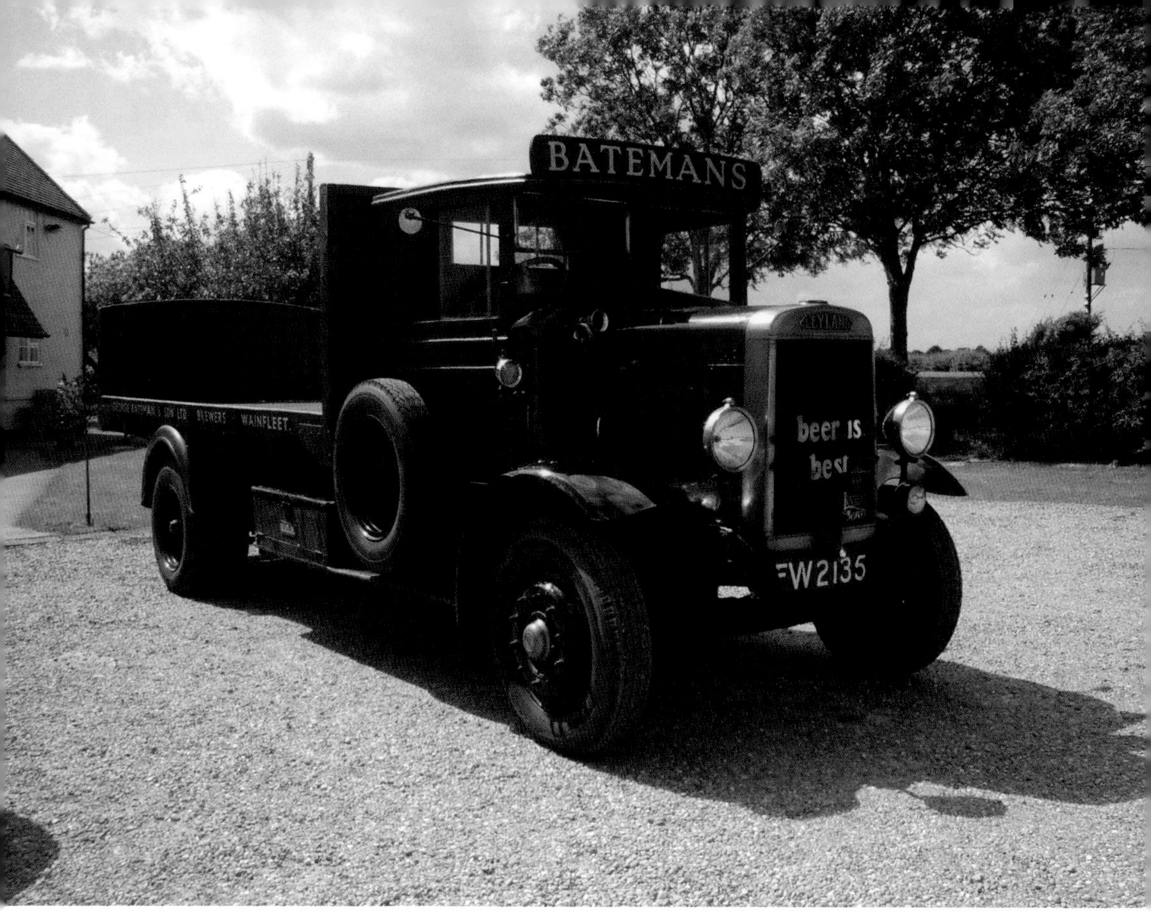

Beer is Best
After retirement from Batemans, this Leyland Beaver dray dating from 1931 was bought privately and restored to its original deep brown livery in near-original condition, complete with the Brewers' Society 'beer is best' slogan. Batemans have since bought a 1965 Austin lorry, which they have restored as a brewery dray and is on display at the brewery. (Batemans)

Until then he'd been using a motorbike, travelling through all weathers in his mackintosh and leggings.

For Harry personally his marriage in 1919 to Charlotte Gunson must have been a joyful event after the hard years of war. Charlotte's father, a prosperous livestock farmer originally from Donington near Boston, had caused something of a stir locally when he'd moved all his animals and family by train to Kent in the early 1900s, despairing of agricultural conditions further north. The family can't have cut all ties to Lincolnshire, however, otherwise it's difficult to see how Harry and Charlotte could have met. There was great sorrow when Harry's father George died in January 1921 at the age of seventy-six, having outlived his wife Susannah by ten years. Over forty-seven years he had steered the business with great care, in his final years advising and helping his son continue the transformation from the tiny brewery he'd founded. Harry and Charlotte moved into the brewery house, where their three children were born. Helen was the first, in 1924, followed by George in 1927 – named after the grandfather who sadly did not live to see him – and finally John in 1929.

The immediate post-war years were not encouraging, with several smaller breweries unable to survive; a major flu pandemic between 1918 and 1919 with some 228,000 deaths in the UK only added to the general gloom. The costs of raw materials had spiralled. 'Everything is more than double the cost of three years ago' complained Harry early in 1919, and his workmen like many others were demanding a shorter working week. His response was to invest in more labour-saving machinery and increase salaries by 4*s* a week in March 1920. This all contributed to 'a most serious loss' in 1920/21, but the business quickly recovered, so that between 1918 and 1921 Harry was able to almost double the size of the tied estate. By the end of the 1920s, Batemans had fifty-seven pubs, including a meaningful presence in local town centres, and two well-regarded hotels, the Peacock and Royal in Boston and the Vine in Skegness, which Harry spent large sums on renovating. Through the decade, however, national beer production was in decline and beer had become more expensive in real terms.

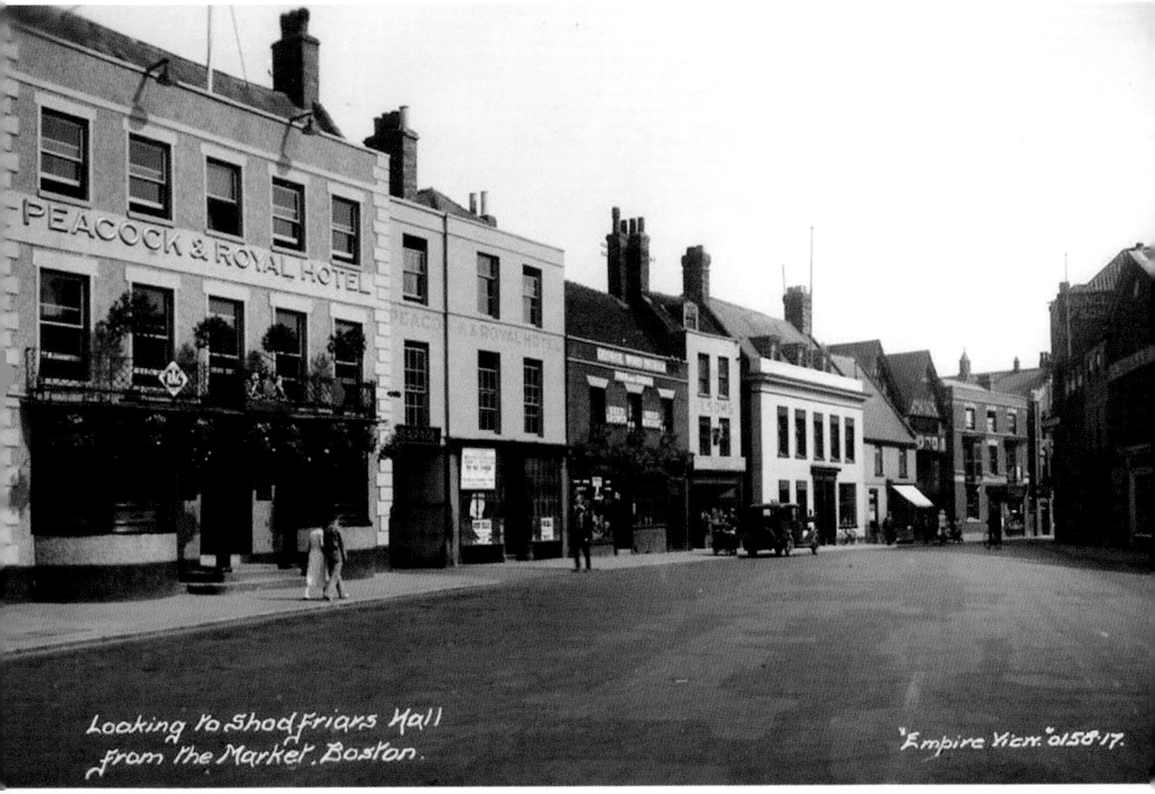

Premier Inn

The Peacock and Royal in Boston marketplace, seen here in a postcard view looking south. Batemans bought the hotel in 1926 and thoroughly renovated it. Judging by the fashions, the single car parked by the roadside and the hotel name in the raised gold lettering that appears on other Batemans pubs of the interwar years, this image was probably taken in the late 1920s. Just visible in the distance on the left is Pilgrim House, the home of Ridlingtons, before the full-length shopfront was added.

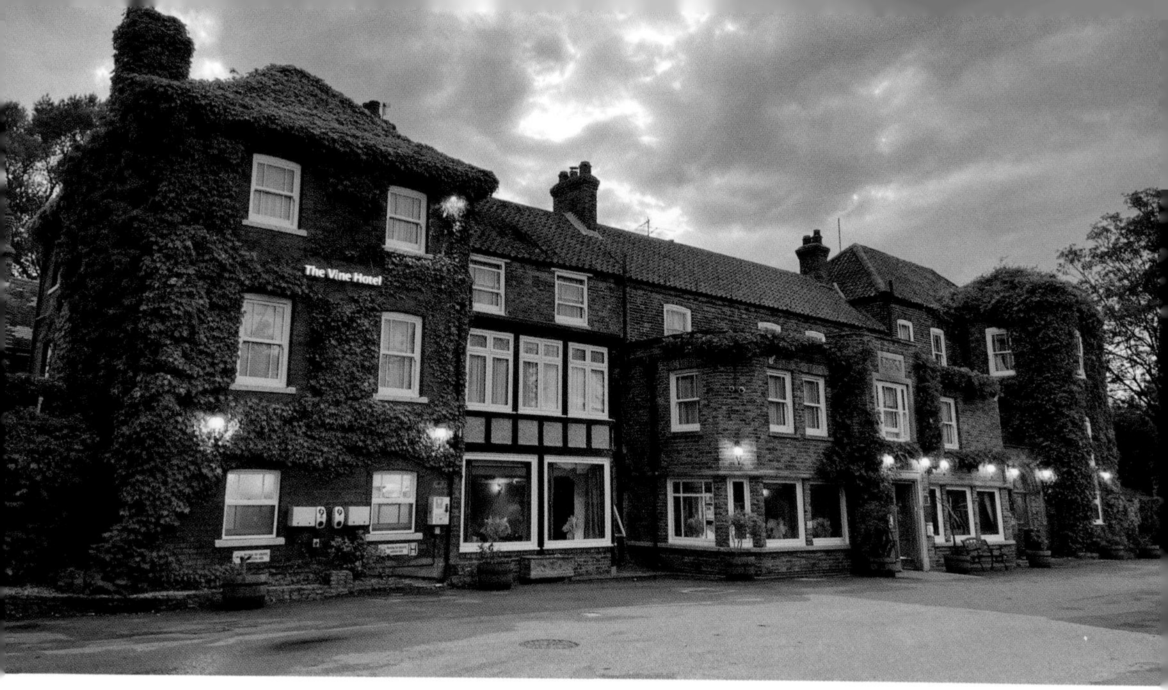

Having a Vine Time
Batemans owned the Vine Hotel in Skegness for many years, although it has now been sold. It's one of the oldest buildings in Skegness, parts of it dating back to the 1770s, though it's much altered and extended since then. During building work in 1902 the skeleton of an exciseman from a century or so ago was found behind some brickwork. The Lincolnshire coast was once notorious for smuggling, and it's rumoured that the officer's ghost still haunts the hotel. (Neil Storey)

How had Harry managed to buck this difficult market? In part he had the cushion of a robust club trade, and because several local breweries had been unable to continue against the severe headwinds, Harry managed to buy several houses that could otherwise have been out of his price range. When Tom Walls, owner of the Burgh le Marsh brewery, a few miles to the north of Wainfleet, retired in 1920, Harry bought the brewery and six of the fifteen houses, some with the help of some post-auction dealing. The next year the Hundleby Brewery, near Spilsby, threw in the towel. It had long been vulnerable to tough market conditions, having expanded too fast with many changes of leadership and buying some mediocre houses. On request from the liquidator, Harry agreed to step in and supply all twenty-three of their houses, and when the estate was put up for auction, he bought five of their pubs. He spent no more than £1,000 per pub around this time apart from splashing out £1,150 on the Britannia in Boston in 1924, ideally sited just off the marketplace with a seven-day beerhouse licence. This is the pub that appears in the iconic view of the Boston Stump and Harry ensured its south-facing wall was fitted with prominent advertising for Batemans Good Honest Ales, which survives to this day.

Other incredibly cheap pubs from this time were the rambling Three Tuns near the coast at Ingoldmells, a major money spinner for many years but now closed, bought for £700 in 1920; the King's Head in the centre of Horncastle, bought for £800 in 1923, and the

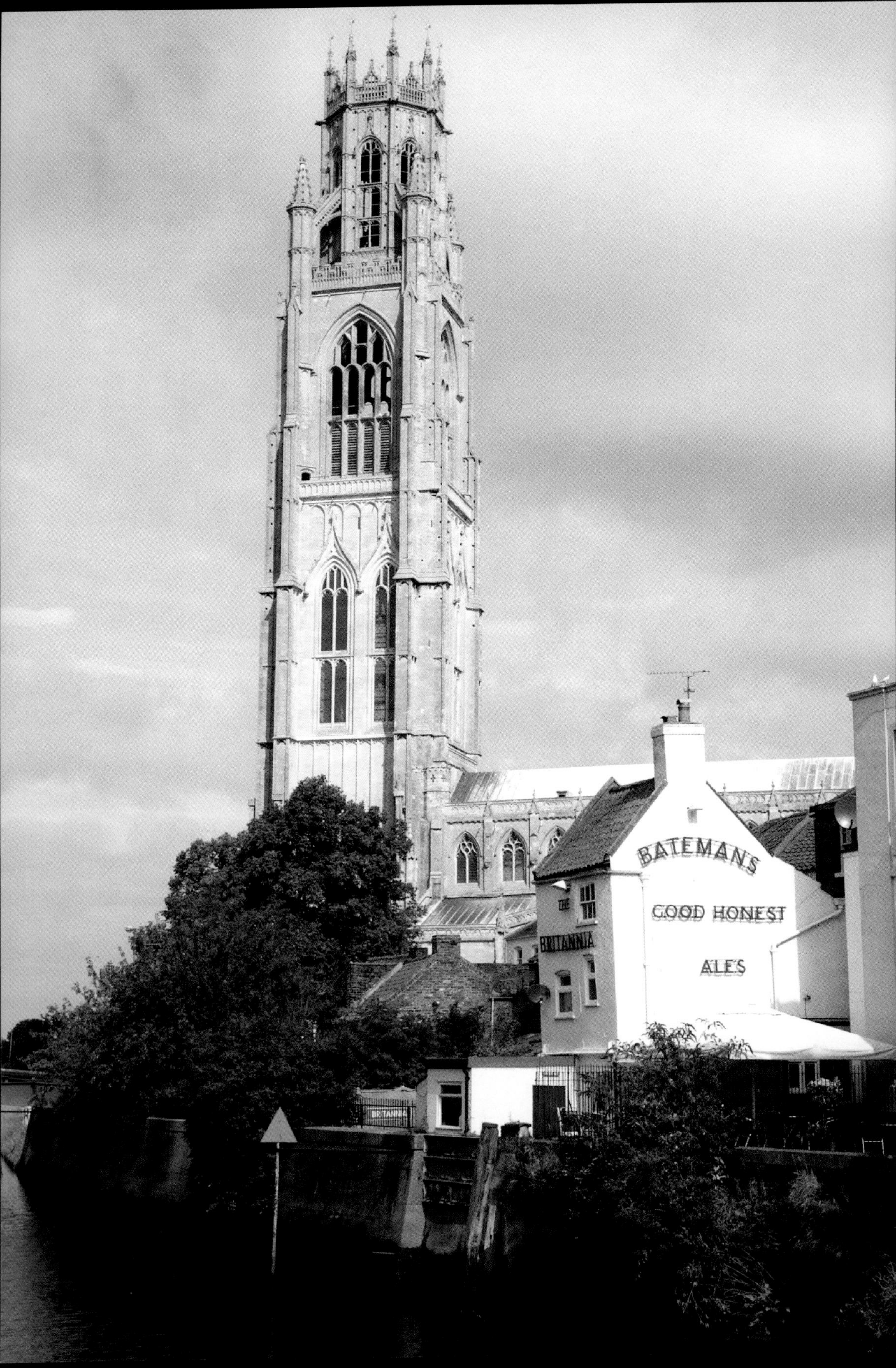
BATEMANS
GOOD HONEST
ALES
THE
BRITANNIA

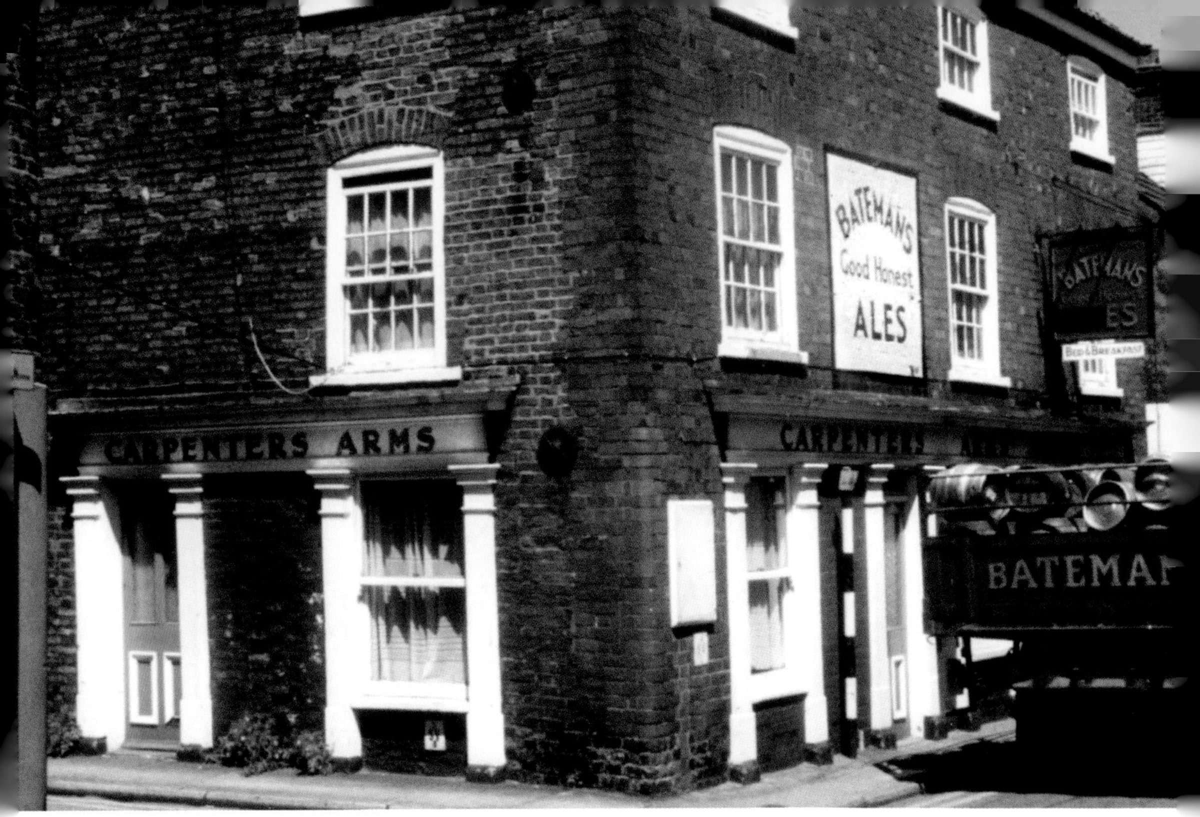

Above: Backstreet Boozer
Much-needed supplies arrive at the Carpenters Arms in Witham Street, Boston, a popular local in the back streets north of the town's marketplace. The image dates from the early 1980s but many features hark back to earlier times, such as the brown livery of the lorry, carrying a jumbled assortment of barrels, and the red pub sign with its jaunty gold letters. The pub name is at the top of the sign, playing second fiddle to the 'BATEMANS Good Honest ALES' strapline underneath, also repeated on the front wall. The house dates from the early nineteenth century and was bought by Batemans in 1928. It remains open today, widely known as 'The Carps', offering live entertainment and accommodation. The front wall now carries the image of a carp wielding a saw. (John Law)

Opposite: Stumped?
This view of Boston's parish church, ironically known as 'The Stump' as the tower is 272 feet high, has featured in many photographs of the town over the years. Harry Bateman realised that and made sure his 'Good Honest Ales' were advertised on the wall of the Britannia pub. (Batemans)

Bricklayers Arms at Old Leake, which Harry finally got his hands on in 1924 for £1,000. The last two are still in the fold today: with its thatched roof, the King's Head is particularly distinctive. The £3,500 that Harry spent for the Duke of York in Boston (now closed and demolished) in 1928 was a rare extravagance. In 1933 Batemans bought the remaining pubs and brewery of the Horncastle and Kirkstead Brewery at auction. Whilst the three pubs were useful houses, the firm's Phoenix Brewery in South Street, Horncastle, was closed and a decaying collection of stock and assets sold off that September. Although Harry is understood to have thought the brewery to be in reasonable condition, he had no need of it and so the buildings were sold to a local dairy firm.

But Harry could not fund these purchases from profits; a key part of his business model had always been borrowing, with almost every acquisition part-funded by a mortgage or loan. The primary funding source was local farmers, many of whom were very rich and had known the Batemans for years; consequently, they felt their money was at minimal risk. The rest was facilitated by an understanding bank manager, with whom George and then Harry had worked for years. The original business overdraft of £1,300 granted in 1906 had expanded

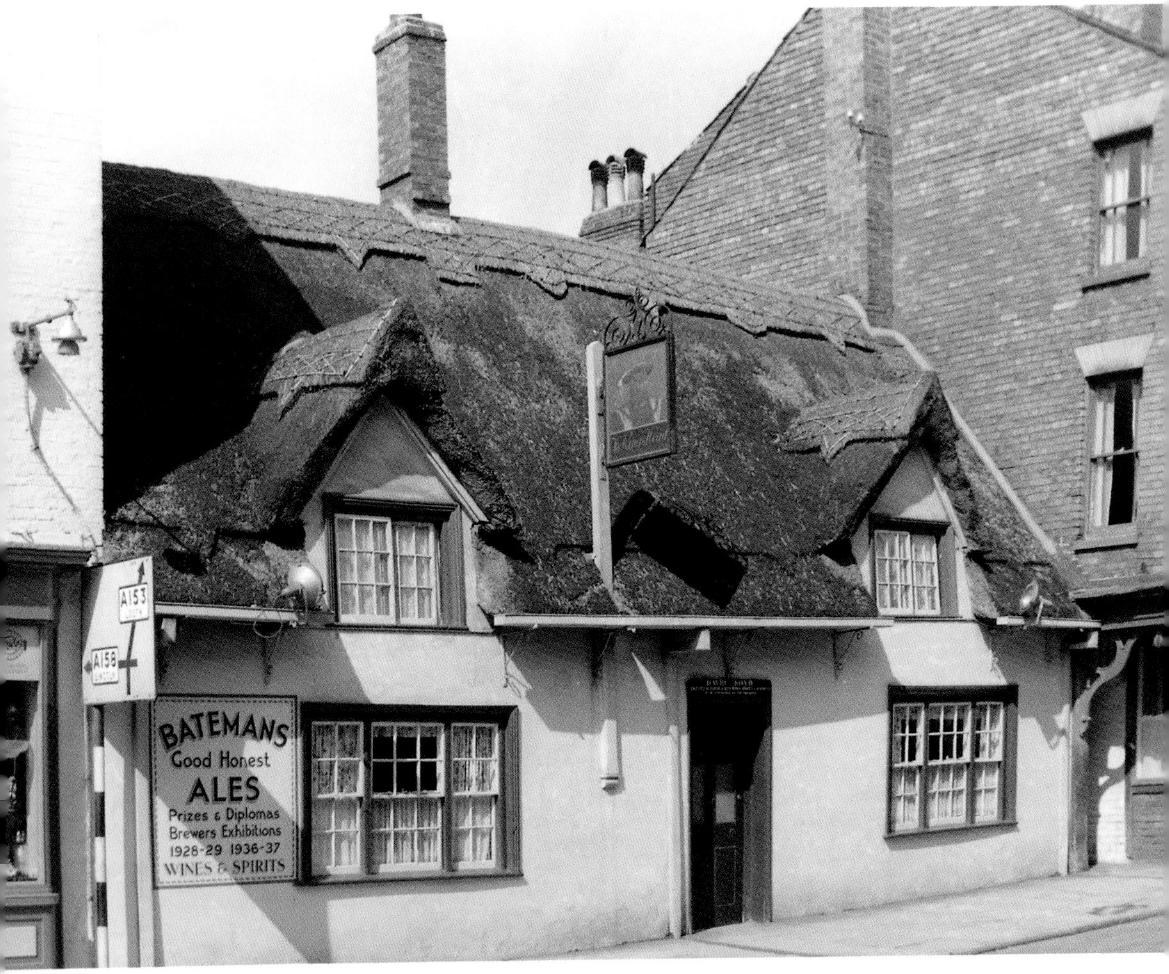

Thatch All Folks
The King's Head in the Bull Ring, Horncastle, is one of two thatched pubs owned by Batemans and was bought in 1923. Its one-and-a-half-storey construction, with 'mud and stud' walling rather than brick, pre-dates most of the Georgian and Victorian town centre; the pub sign 'growing' through the roof is surely unique! David Boyd, whose name appears above the door in this image, was landlord for around twenty years until he retired in 1969. The house, like many others in the vicinity, was damaged by extensive floods in 1960, which according to Mr Boyd saw the end of the pub's resident ghost! (Batemans)

to £18,000 by 1920, although when the bank eventually put their foot firmly down, Harry managed to reduce the debt somewhat. He disclosed to the bank that he'd considered taking in a partner, or forming a limited company, but some business friends had advised him against. Finally in July 1928 the whole business, including Ridlingtons, was incorporated as George Bateman and Son, Ltd. Harry was director and chairman. By September 1929 the issued share capital in ordinary and preference shares stood at just over £49,152, which covered – just – the company's mortgages and charges. Nearly all the investors were local farmers and traders.

It was around this time that Batemans bought a rather different property, the Salem Bridge windmill, whose stylised image has been a key part of the company's logo for years. This six-storey brick-built mill stands across the road from the brewery and was not originally part of the site, having been built around the 1820s, probably by the millwrights Oxley of Alford, where there is a mill to a very similar design. It was used as a flour mill by bakers Charles Handsley and then William Fretwell, who retired in 1909 and passed the business on to his son, also called William. Unfortunately, Fretwell Jr was declared bankrupt five years later owing over £1,500, then a significant sum. It seems unlikely that the mill worked beyond December 1913, when Fretwell's business tools and stocks were sold at auction. Harry was at first interested in acquiring the mill to carry on his existing bakery trade, but agreeing a price with Fretwell Sr dragged on for years, and it was not until 1920 that the sale took place. By this time Harry was more interested in using the mill and its outbuildings as a bottling store, which explains why he was prepared to pay £1,142 for it – at the time, he was paying less than that for most of his pubs. The mill's machinery, its five sails and cap were dismantled and sold the following year, and replaced with a protective roof. Mrs Bennett said:

> They decided what an advert it would be to put a bottle weathervane on top of the Mill. Well, poor old Joe Turner [a friend of Harry's] took the job on, oh dearie me, he must have spent all week up and down the mill, he couldn't get it to balance so that the wind would move it. He used to come into the office absolutely exhausted, the tears would run down his face because he was so frustrated. Eventually he got it right!

Advertising was becoming increasingly important in the brewing trade. By the 1920s brewers were realising the importance of a distinctive feature and strapline that customers would remember. Probably the most famous slogan was 'Guinness is good for you', accompanied often with an impressively beaked toucan, first run in 1928; it was devised by the novelist Dorothy L. Sayers, then working as an advertising copywriter. Most campaigns, though, tended to look back to an imagined 'Olde England' with beaming huntsmen or avuncular 'mine hosts' in traditional surroundings. Batemans came up with their own rather better slogan, not accompanied by banal cartoons: 'Good Honest Ales' saw the light of day in local newspapers during July 1923, from an idea by Mrs Bennett: 'Mr Jameson [the Managing Clerk, who left in 1926] wanted "Batemans' Good Ales" and I said that's not enough it should be "Batemans' Good Honest Ales". There isn't a more honest man than Mr Bateman.' She wasn't to know that a Halifax brewer had used the phrase 'the same good honest ales' for a few months during 1915! Batemans quickly trade-marked 'Good Honest Ales', which was used to great effect for many years (though still a registered trade mark of Batemans, it's now been replaced by more modern advertising appealing to a wider range of drinkers). It was probably around this time that Batemans settled on a distinctive rich

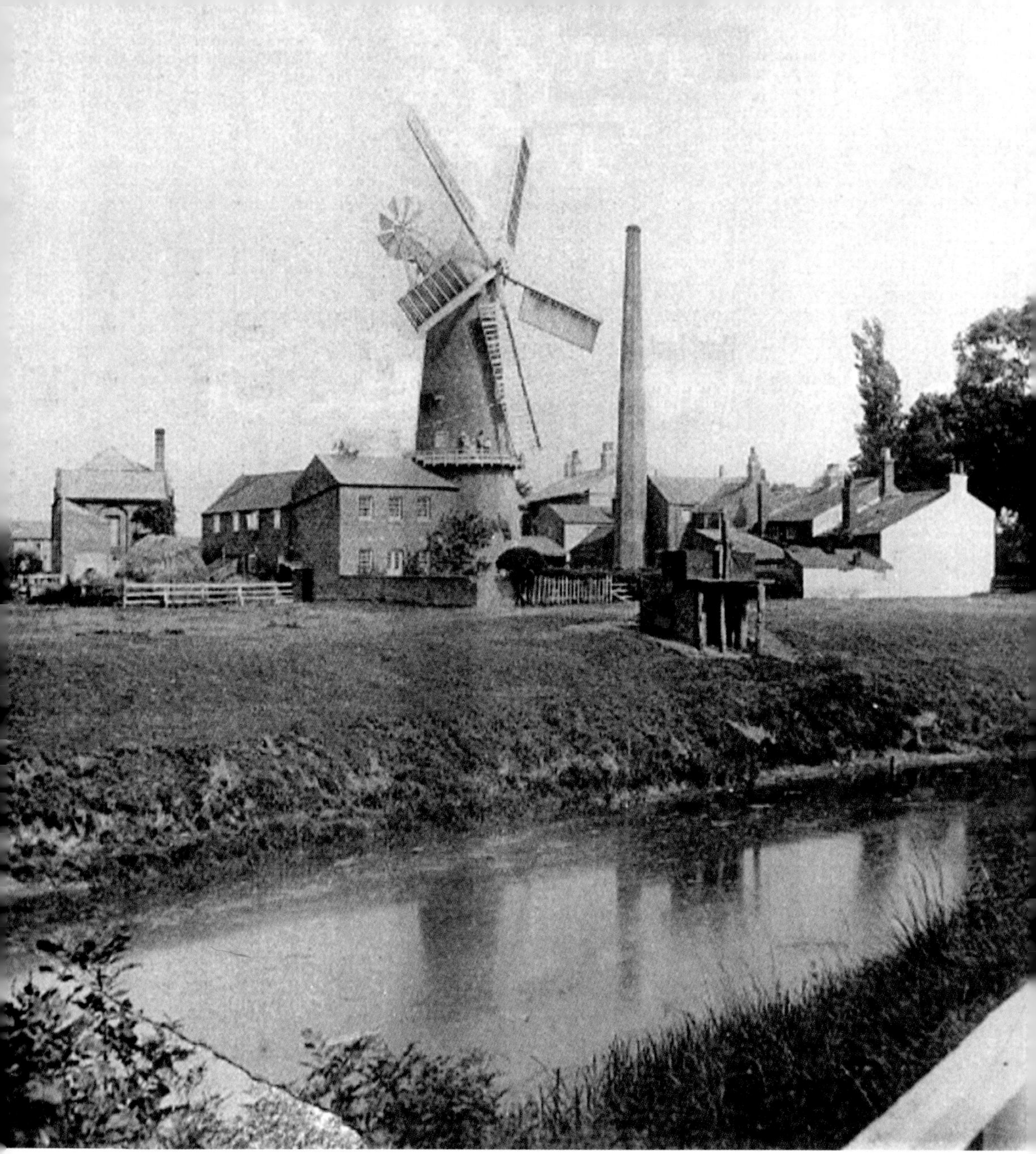

Above: Full Sail
The five-sailed Salem Bridge windmill in its prime around 1900, seen looking north from the Steeping River, with the brewery to the left and the mill cottages to the right. The mill tower today is about the same height, although the main sails, fantail, onion cap and external walkway were all removed in the 1920s when Batemans adapted the mill for their own use. (Batemans)

Opposite: Iconic Image
It couldn't be anywhere else, could it? The ivy-clad mill is seen with the brewery behind it in this delightful photograph. (Batemans)

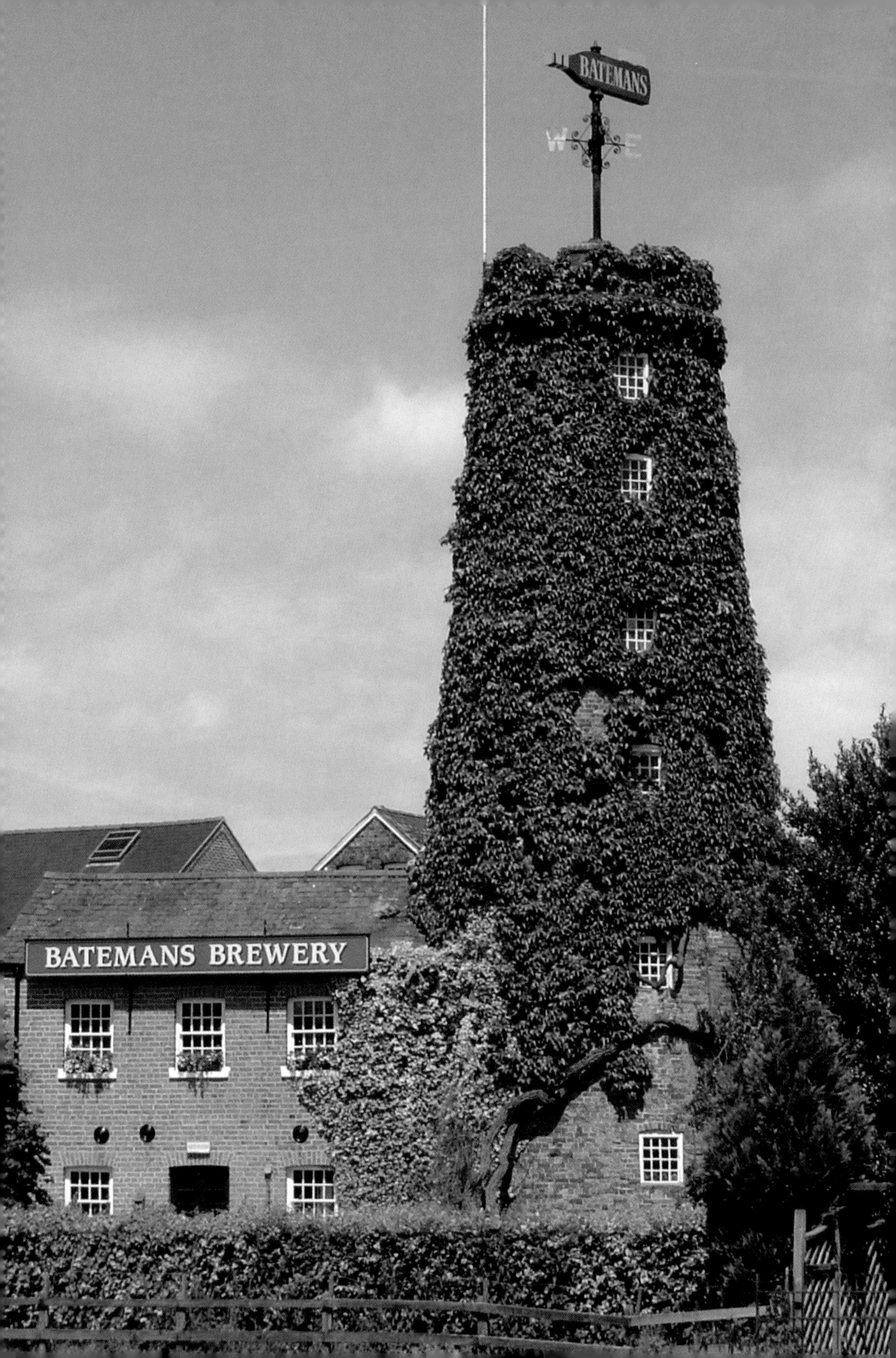
BATEMANS
W
E
BATEMANS BREWERY

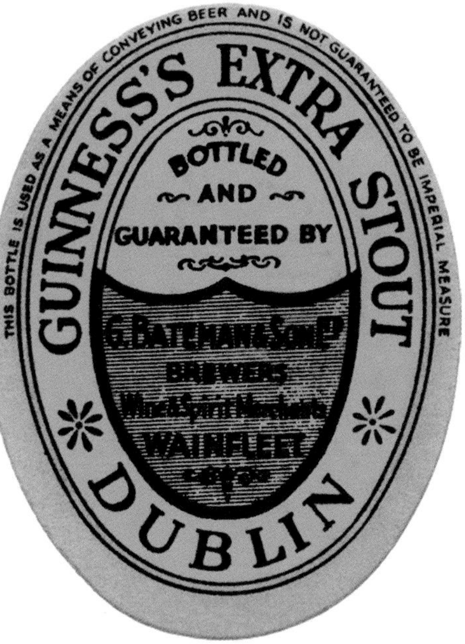

Left: Guinness' Extra Stout
Like many other local brewers, Batemans bottled national brands such as Bass, Watneys stout and Guinness. The latter were delivered in hogsheads (54-gallon wooden barrels) to Wainfleet railway station, all the way from Dublin, and matured for two weeks before bottling.

Below: Double Signage
The Mill at Boston boasted not one but two signs for many years, as shown here in 1980: on the right, a traditional red and gold Batemans sign, the pub name as usual subsidiary to 'Good Honest Ales', and on the left the Batemans wavy logo. (John Law)

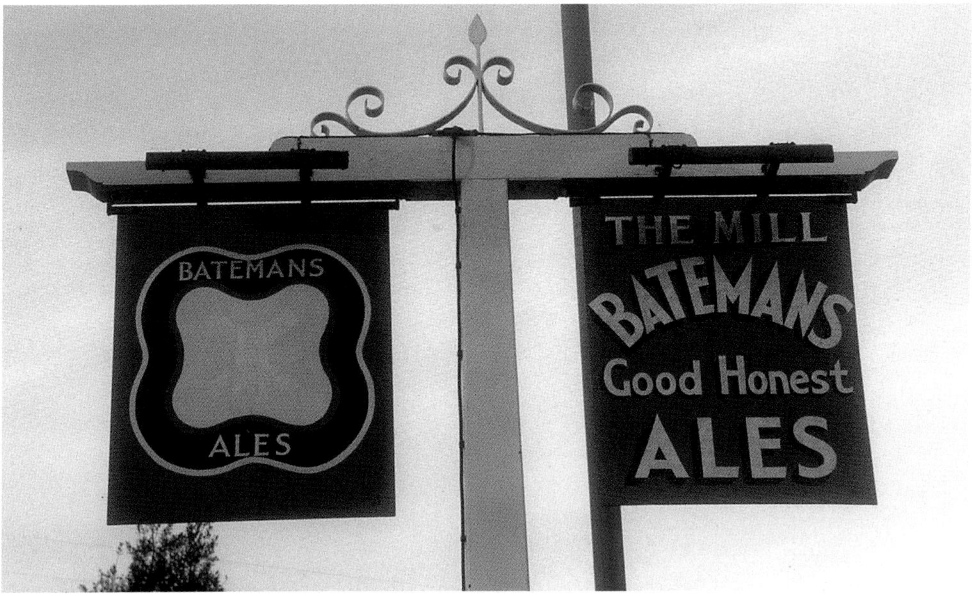

maroon livery, with gold lettering in a jaunty period font, for their pub signs; the firm's lorries and vans were also painted maroon, or sometimes brown. This style, which lasted into the 1980s, made Batemans pubs very recognisable, easily distinguishing them from competitors like Soulbys of Alford (white lettering on green) and Soames of Spalding (half blue and red). All three preferred lettered rather than pictorial signs, a feature which Batemans still follows for most of their pubs.

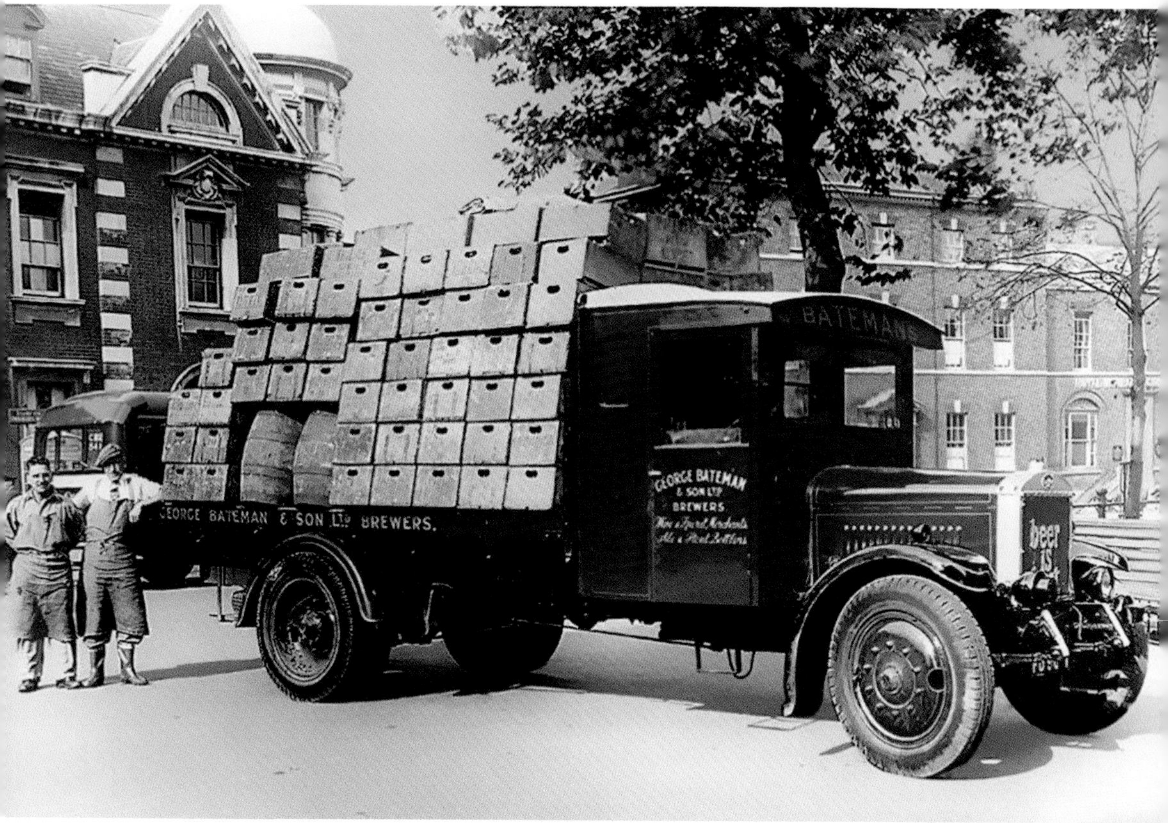

Heavy Load
Two Batemans draymen pose with their elderly Guy delivery lorry outside the General Post Office in Bargate, Boston, in this colourised image dating from around 1945. The lorry dates from the late 1920s and but for the war would probably have been replaced, although it seems well cared for. The offside headlight retains wartime masking, designed to show a minimum of light. The load of mainly bottled beer in wooden crates looks precarious but is neatly packed, its weight distributed inwards, no doubt borne out of experience. A 'Beer is Best' slogan, devised by the Brewers' Society in 1933 to help arrest declining beer sales, appears on the lorry's radiator. In the background, a Wright's of Louth bus waits to leave. (Batemans)

A small but fascinating notebook has survived from the early 1930s, full of short and pithy comments about the tied estate. Probably written either by the head brewer or traveller, the book records roughly monthly visits to each house – or weekly if things were really bad. For most of the time the landlords are doing their best, but trade is often disappointing; some of them are clearly in arrears with their rent, and Batemans are trying to help them with repayment plans. We encounter a range of opinions from the landlords: at the Bull Hotel in Spilsby electric lighting is requested, yet at Boston the tenant of the Duke of York resists such progress as he thinks the bills will be too high. But he is not popular with the unknown diarist: 'I'm afraid he is lazy and careless,' he comments in September 1934, which is downgraded to 'hopeless' a few months later. The tenants at the Foresters Arms in Mareham le Fen are no better, always short of stock and not planning forward. Trade at the Old Chequers, Croft, is 'very poor', although it remains a Batemans house to this day. At times there were good tenants in bad houses, such as the isolated Packet Inn near Tumby Woodside ('a bad old property ... in a dreadful state'), but the beers were excellent despite the lack of trade, which probably discouraged outlay on improvements.

The chill winds of the 1930s recession severely affected most people and businesses, especially heavy industry in the North and Midlands. In Lincolnshire several engineering firms reduced their workforces considerably or in some cases entered insolvency, which must have had a knock-on effect for the drink trade. Batemans added only ten more pubs to their estate in this decade but the impact on their club trade was probably even more severe. Their financial situation was not particularly good either: the bank debt had soared to £50,068 by June 1936. This may have led to the story from this time about Harry initially laying off most of his workforce, only to take them back on later when he saw them on the streets of Wainfleet. Harry may also have needed his staff back because he had spotted new opportunities. The first was supplying the expanding holiday empire of the entrepreneur Billy Butlin, a showman very much in Harry's mould. Butlin had opened an amusement park in Skegness back in 1927, and in 1936 established his first holiday camp at a vast site in Ingoldmells, just north of Skegness. Batemans already had a large pub in Ingoldmells village, the Three Tuns, which as a result of the new camp quickly became one of their best-selling outlets, and the opening of Butlin's Ingoldmells Hotel in June 1939 also helped trade. Harry and Billy Butlin had worked together on various local committees and got on well; it was little surprise when Batemans were contracted to supply beer to the Butlin holiday empire.

The second opportunity arose from the 1930s trend for pub improvements, whereby some of the larger brewers either substantially rebuilt an older pub, or demolished one and replaced it with something much more appealing to the eye, at least to the police and magistrates. Others constructed large new pubs, often on new housing estates or development sites. Harry kept a close eye on proceedings at the old Ship Hotel in Skegness, an undistinguished but large Victorian pile sitting at a key road junction between Roman Bank and Castleton Boulevard. The owners, Nottingham-based Home Brewery, were expanding across the East Midlands and had obtained permission for a replacement on the seaward side of the junction, designed in a then trendy 'moderne' style, all flat roofs, reinforced concrete and metal-framed windows. The new Ship opened in April 1934 (and survives today under a Grade II listing), but Harry had his own plans: rather than a mere pub, he would build Skegness a new modern hotel right on the seafront. His first thought had been to call it the 'Jolly Fisherman', after the cartoonish character who leaps along the vast golden beaches of Skegness, first made popular in John Hassall's poster for the Great Northern Railway. But he rejected that idea for a more upmarket and ambitious name: the County Hotel. In a nice touch, the foundation stone of the twenty-nine-bedroom hotel was laid in September 1934 by Harry's children, Helen, George and John, then aged nine, six and five. Opened in June 1935, it even sported a roof garden, a lounge bar designed to resemble a Spanish galleon, and 'beautifully fitted toilet accommodation' in every room. Harry could not resist some references to his original idea: either side of the fireplace in the cocktail bar there was a coloured glass representation of not just the Jolly Fisherman, but a Jolly Fisherwoman as well! The expense of the County, and the Vine Hotel further

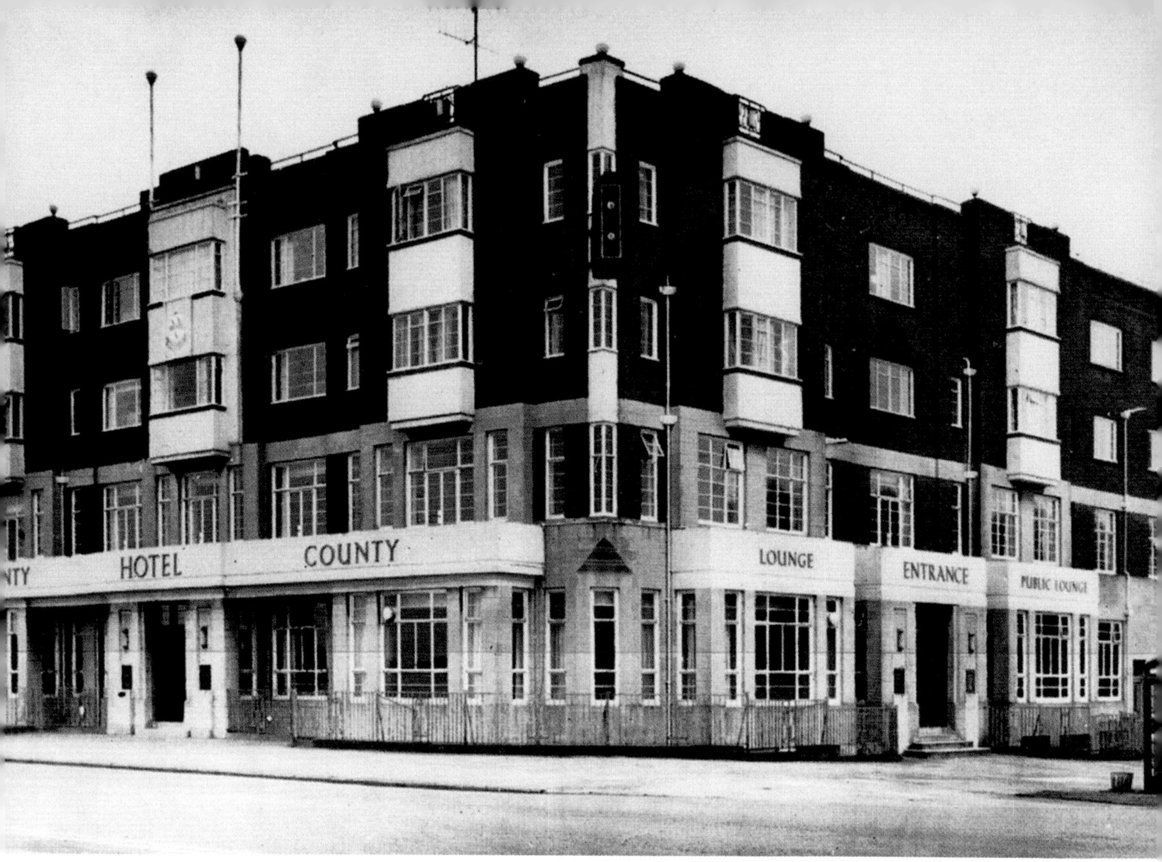

Seaside Special
Despite the severe economic depression of the 1930s, Harry Bateman pressed ahead with building the County Hotel on Skegness seafront. It opened in April 1934 and is still in use as a hotel, though Batemans sold it some years ago. Its four floors tower above the surrounding properties, although it's now dwarfed by the five-storey, eighty-room Premier Inn recently erected on the foreshore closer to the Clock Tower.

south, both of which were sold in the 1980s, is the main reason why Batemans now has no meaningful presence in Skegness, despite the resort being only 5 miles from the brewery.

Batemans ended the decade in a more stable state. The brewery was producing nearly 10,000 barrels of beer a year (surpassing several much longer-established businesses), the tied estate had grown to sixty-seven pubs and hotels, there was a healthy free trade, in Ridlingtons Harry had a flourishing wine and spirits business, and Batemans were still supplying many of the farming families who had been their customers for years. Harry began to be recognised, too, as a key player in the brewing industry; he became chairman of the East Midlands Brewers' Association in 1935, representing the interests of his members across a wide region, nearly all of whose businesses produced far more beer and had a greater number of pubs. Harry established lifelong friendships with many of his fellow brewers. Firms not only sought his advice: two local brewers, Mowbrays of Grantham and Soames of Spalding, whose directors were mostly very long in the tooth, went so far as to invite Harry onto their company boards as a director, such was the trust placed in him. This respect was not confined to the brewing industry; Harry became a director of the Lincolnshire Standard newspaper group in 1944, and was also active in a number of local interest groups.

A new head brewer, Montague Broad, had joined the company in 1926 from the Stratford (London) brewery of Savill Brothers, which had just been bought and closed by Charringtons. Originally trained as a brewer in Lewes, Sussex, his brother was also in the brewing trade, working for Eldridge Pope at Dorchester. Broad was a widower with two young daughters, his wife having died in her early forties. He spent the rest of his career – twenty-six years – with Batemans, retiring in 1952. He carried on as a director however, living at Mill Cottage until his death in 1974 at the ripe old age of ninety-three. Chartered accountant William Bell joined in 1936, also becoming a director of the company. Hugh Jackson started in 1940 and over the years rose to be transport foreman. The record for length of service, which has yet to be beaten, is fifty-one years, held by Harry Waugh, who began with the firm in 1924 at the age of thirteen and retired in 1975 as foreman of the bottling store – despite his role being mainly a manual one, he would always come to work

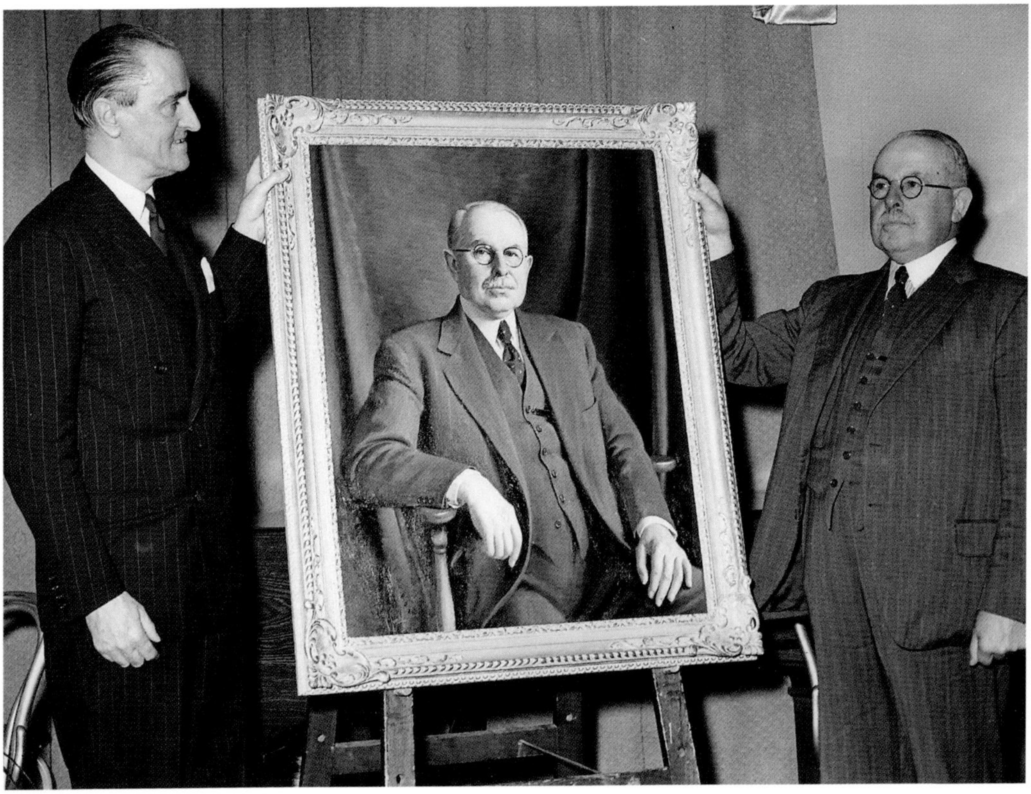

Mirror Image
On his retirement as chairman of the East Midlands Brewers' Association in November 1953, Harry Bateman was presented with his portrait in oils by his successor, William Hanson of the Nottingham brewers Hardy & Hanson. Here they both are, posing rather stiffly; Harry seems to be wearing the same three-piece pinstripe suit and tie. The artist was Carlos Sancha (1920–2001), a Spaniard who served in the RAF during the war and lived locally for several years. (Batemans)

in a pristine suit and tie. These long-serving workers were the backbone of the business as it again hit difficult times when the country was once more plunged into war. Conscription became a problem again, with all able-bodied men aged between eighteen and forty-one (later raised to fifty-one) liable to be called up. Several of the younger brewery staff served in the armed forces all over the world.

Despite the similarities around conscription, the Second World War was perhaps not quite so drastic for the brewing industry, although as ever the impact differed up and down the country. Although there were some restrictions on production, and beer duty inevitably soared again, the imposition of some form of nationalisation on the trade was far less of a worry. Lincolnshire was seen as a key base for siting bomber squadrons, especially on the undulating uplands of the Wolds and Lincoln Edge. Several Batemans pubs became favourite watering holes for bomber crews, both RAF and USAF, many of whom sadly failed to return. Pubs like the Royal Oak at Mareham le Fen with its thatched roof must have seemed 'typically English' to the American crews of Flying Fortresses. The Red Lion at East Kirkby (now a free house) has a table with the initials of bomber crews scratched into it, and a vast amount of wartime memorabilia.

Harry was keen to 'do his bit' for the war effort. At first he helped to administer petrol rationing locally, but then events took an unexpected turn. An approach was made in

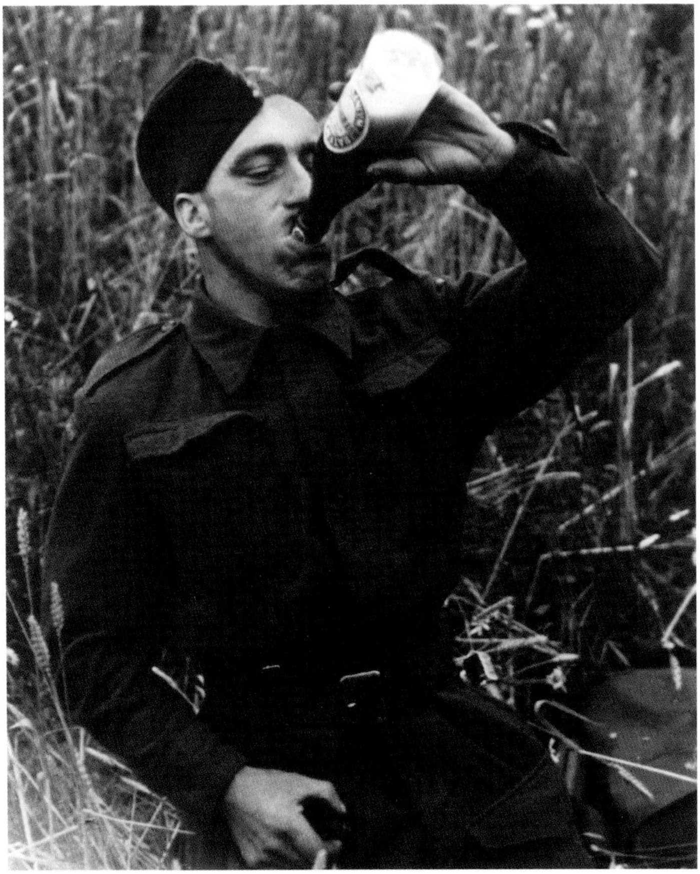

Hop to it, Private
This is probably a staged photograph, but it still draws the eye. A Second World War soldier refreshes his thirst with a bottled pint of Batemans Nut Brown, a bitter ale that was brewed until the late 1980s. The yellow and blue bottle label would be highly prized by collectors these days! (Batemans)

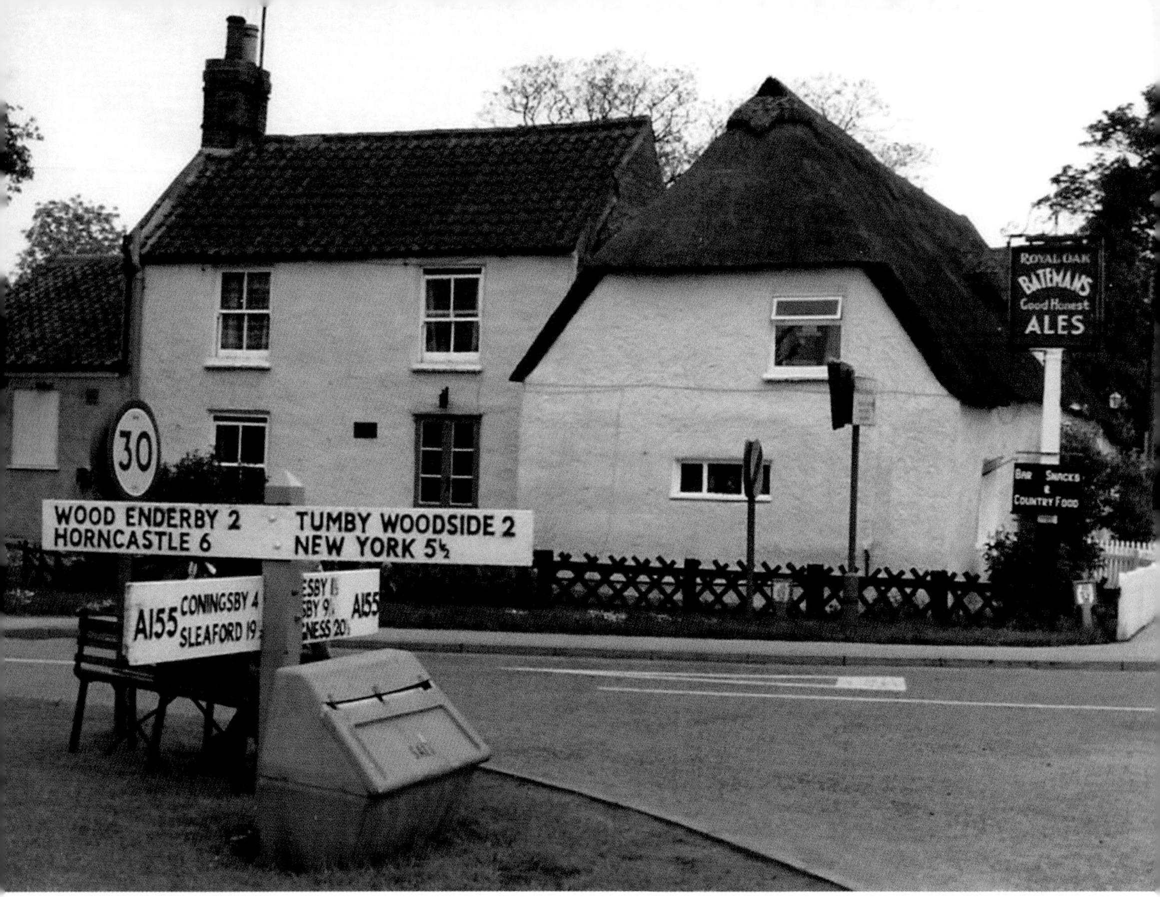

Essence of Lincolnshire
The Royal Oak at Mareham le Fen is a 'mud and stud' building dating from the seventeenth century, with later additions, and is one of the oldest thatched pubs in Lincolnshire, if not England. It's been a Batemans pub, apart from a brief spell as a free house a few years ago, since 1957. The road sign, with its typical Lincolnshire village names, is a gem. (In Search of Susan blog, susansearch.blogspot.com)

great secrecy which led to Harry joining a clandestine wartime resistance organisation – a civilian M16, in effect. The Auxiliary Units – a suitably anonymous cover name – were divided into two divisions: an operational resistance, devoted to disruption of invasion forces by any means possible, and a signals unit, given the codename Special Duties Section. This was a network of local wireless networks overseen by the Royal Corps of Signals, with outstations near the coast, but largely staffed by civilian radio operators and observers. Whilst some secrecy still surrounds these operations, it's thought that over 3,000 men and women were involved across local communities. Those asked to join were often key local figures such as bank managers, doctors, vicars – and in Harry's case, brewery owners. The Lincolnshire network was based at Canwick Hall near Lincoln and had ten substations, including Harry's, all operating under the codename 'Ormskirk'. Wireless sets and their aerials were very well hidden; it was all run on a very strict 'need to know' basis, with all civilians involved required to sign the Official Secrets Act. Indeed, secrecy was such that it's believed most members of the unit did not know who the others were; many years after the war one of the Canwick Hall military staff recalled that he spoke to Harry by radio many times without ever knowing his name. There was an important point to these precautions: the attitude of German occupying forces to resistance organisations was well known. If an invasion had taken place, any captured Special Duties personnel would very probably have been shot, so people like Harry were taking significant risks. The Wainfleet substation was set up in a garden shed at Bridge House, then Harry's home, with the aerial running up the trunk of a copper beech tree. Harry, along with most other civilians, was stood down from his military duties in 1944, although his cellar still contained unwanted transmitters, anti-tank guns, grenades, machine guns and other military hardware for many years afterwards. In 1953 he was awarded an OBE 'for political and public services in Lincolnshire', although his wartime service was no doubt also a factor.

Batemans were very lucky that no serious damage was sustained to any of their pubs or the brewery, despite their presence on the tempting east coast – some Luftwaffe bomber crews would drop any remaining munitions over Lincolnshire as they headed home, although most raids over the county were targeted at industrial and strategic sites around

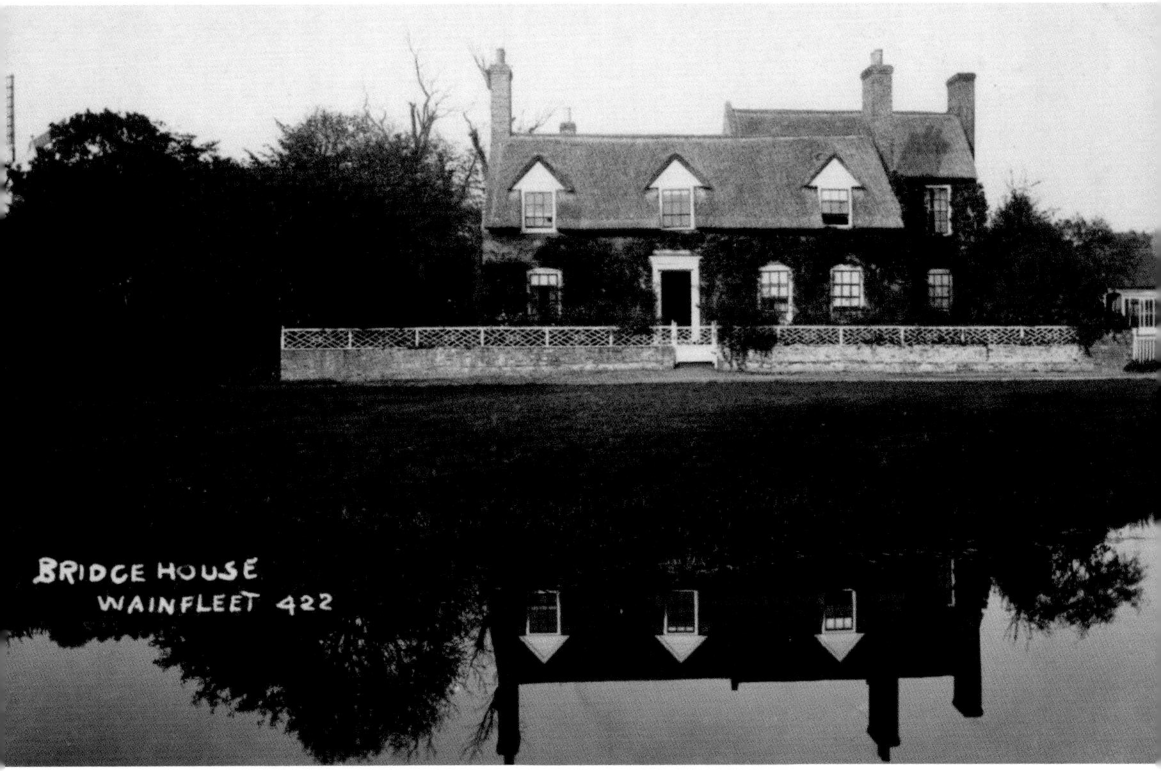

Riverside Home
As well as Salem House, Batemans also own Bridge House, just around a bend in the river on the opposite bank. It has been the home of successive brewery proprietors over the years and was the location for secret defence operations conducted by Harry Bateman in the Second World War.

Grantham, Grimsby, Scunthorpe and Lincoln. Wainfleet did not escape damage: a German bomber dropped three high-explosive bombs on the town in November 1940, leaving a baker and his wife dead and causing significant damage to The Vinery next door, the former home of the Gunson brewing family (it was later demolished and is now the site of the town's Co-op). The physical impact on Batemans' property was limited to some minor damage at the County Hotel in Skegness, and a machine gun attack on the brewery itself by a returning German bomber – the old bottling store roof still has the bullet holes. The effect of the war on those workers who had been called up, though, was tragic, with one member of staff dying in a prisoner of war camp, and no doubt others were left with difficult memories of wartime service.

A little-realised side effect of the air war in the county was the hiring of rooms at local pubs on a supposedly temporary basis to house crews and ground staff as quickly as possible, where barrack construction was not keeping pace with demand. This practice started in 1940 with the hiring of rooms at the Nelson Butt and Bull Inn at Spilsby, and quickly expanded through 1941 and 1942 as existing airfields were improved, concrete runways replacing grass in many cases, and more land taken over for further airfields. Pub bedrooms and clubrooms were often requisitioned, then de-requisitioned and then taken up again, which was no doubt unavoidable but presented landlords and Batemans with a very unpredictable situation. The Wheel Inn at Welton le Marsh, for example, had seven bedrooms requisitioned in January 1942, probably in connection with the construction of RAF bases at Strubby and Spilsby, having already been requisitioned and then de-requisitioned in 1940. For some other pubs, it was just the one room that was needed, which was fortunate in the case of a very small beerhouse, the Fox and Hounds at Hundleby, near Spilsby. Even the brewery field at Wainfleet was found a military use, perhaps for parking vehicles and equipment. Some of the pub space was not relinquished until 1946, and claims for repainting and repairs dragged on into the following year.

6

Striking Out Alone, 1945–85

Once again, the immediate post-war years were not especially good for Lincolnshire brewers, although this time much of the damage was self-inflicted. Harry must have felt disappointed, and in some cases annoyed, at what took place. Within the space of less than seven years, what had been a largely harmonious collection of brewers mostly working with each other to meet market demand had been badly damaged by the sale of three of the biggest firms to larger regional brewers remote from the county – and Harry was on the boards of two who opted to sell out. The first to go, Soames of Spalding, was probably inevitable, with a mostly ancient and creaking board of directors, worn out by wartime demands, anxious to retire. They sold their brewery and 240 pubs to Steward & Patteson of Norwich in 1949, and Harry resigned after five years as a director. The next steps were less amicable. The brewers J. W. Green were rapidly expanding, snapping up breweries all over the country, many of them very remote from their Luton base. Their attentions fell on two Lincolnshire breweries: Soulbys of Alford and Mowbrays of Grantham, who between them controlled some 350 pubs. The Alford firm was bought by Greens in September 1951, who almost immediately closed the brewery. Soulbys was slightly smaller that Mowbrays, with a similar pub estate to Batemans across much the same area; the Grantham brewers had a greater presence in larger towns and served a wider district. Harry was on the board of Mowbrays, but not Soulbys.

Very much like his role at Soames, Harry had visited many of the pubs owned by Mowbrays and had a good grasp of the business. It was more profitable than Soulbys, it had raised more cash by the issue of further shares in 1948 (Harry bought some of them) and probably was a better proposition all round. It was therefore a shock to Harry when he arrived for a board meeting in December 1951 to find that discussions were taking place with Greens regarding a takeover. Whether he had been given assurances that a sale would not happen is not known, but he resigned by telegram seven days later, Mowbray's chairman referring to 'personal matters between Mr Bateman and himself'. The purchase of Mowbrays went through very quickly, and though brewing at Grantham carried on for a short while, beer was soon being hauled up to Lincolnshire from Greens' main breweries in Luton and Stratford on Avon.

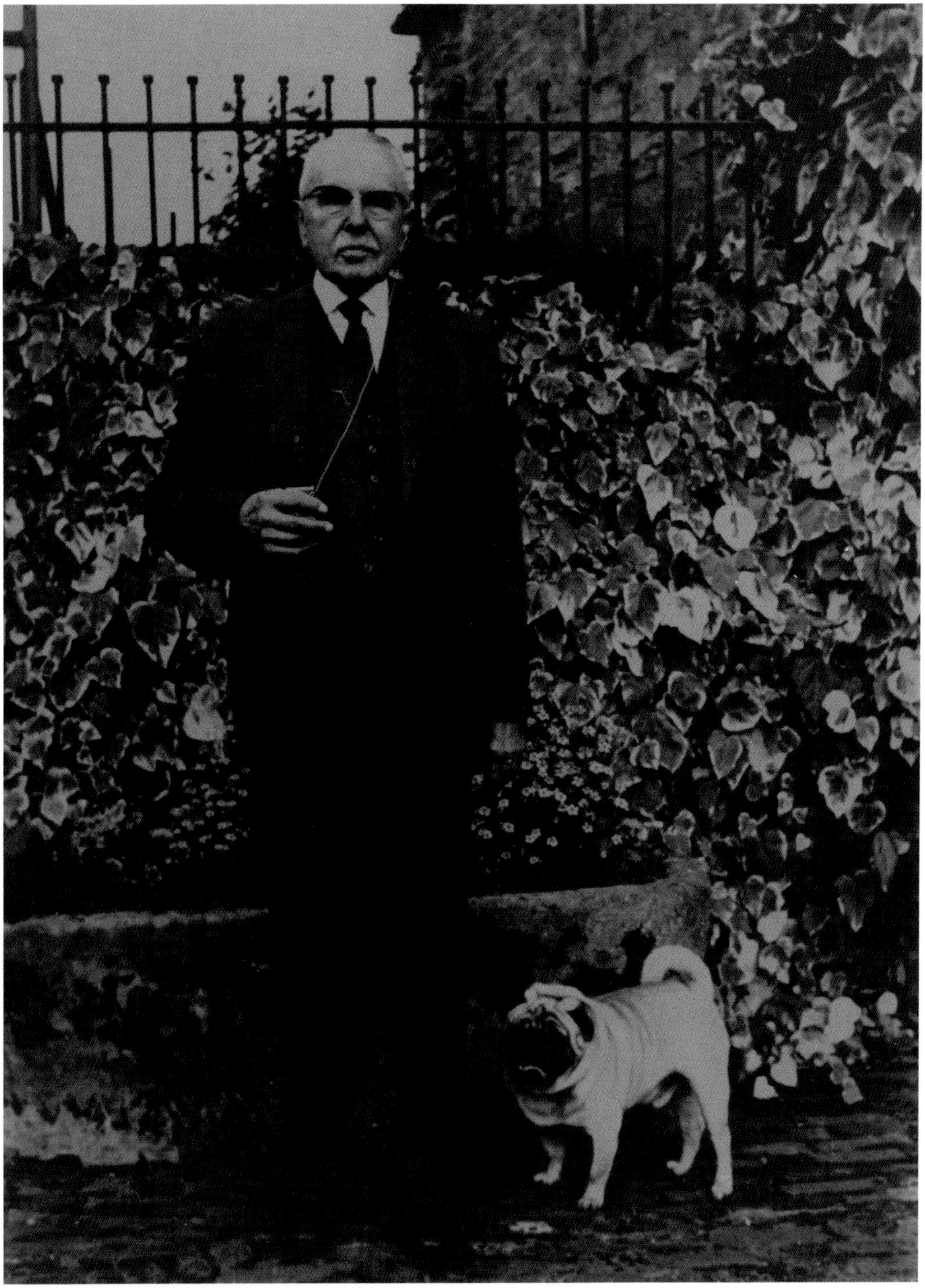

Bulldog Spirit
Harry Bateman pictured in later life, perhaps around 1950, in the garden of Brewery House with his pet dog. By this time he was quite deaf, hence the hearing aid he is holding. (Batemans)

Batemans, with their seventy or so pubs, were now surrounded by much bigger firms; as well as Greens and Steward & Patteson, there were five Nottinghamshire brewers who between them owned about 500 houses, with Hewitts of Grimsby and Halls of Ely also owning large pub estates in the county. But Harry was undaunted; owing to Batemans' modest size and perceived geographical remoteness they were an unlikely target for takeover, although many years later Harry's son George remembered that Greens had contacted Batemans 'with the idea of forming a Lincolnshire brewing group. We said no.' He also suggested Batemans did not fancy being the junior partner in a much bigger concern, but it seems just as likely that Batemans valued their independence and the jobs of their staff highly, as later events were to prove. Batemans had been investing heavily in the brewery, starting with a new hop back and refrigeration equipment in 1947, followed the next year by a replacement copper, bought second-hand from Tetley's of Leeds for £500. A new boiler from the Lincoln engineers Robeys was installed in 1951, costing £5,000, and a new bottle washing machine was added. The water supply, too, had been improved; for several years water had to be transported 7 miles by lorry from the nearest water main, but at last in 1947 reliable mains water, suitable for brewing, reached Wainfleet.

The biggest changes though came in 1952 with the total transformation of the fermenting room. Harry's son George – who joined Batemans as executive director in 1950, after training, rather ironically, with those merger addicts Greens of Luton – advised his father to end the carriage cask system. This was an antiquated hangover from earlier times by which the wort (the liquid produced in the mashing process) was transferred into casks positioned over pine troughs in ground-floor rooms; the casks had to be topped up with a can from the trough every two hours, day and night, and when this tedious task was complete the beer was then sent to pubs in the same cask. Not only was this hugely labour intensive, about 8 per cent of the liquid could be lost and in the summer months the heat generated was almost unbearable, even after George had the roof slates painted white to try to reduce the temperature. It was a miracle that the beer was actually drinkable, George claimed years later. The investment needed to end this onerous practice was significant at a cost of £13,800: eight new stainless-steel fermenting vessels were installed along with an air-conditioning system to keep the temperature down to 15° C (as opposed to the 30° that could previously be achieved). This increased brewing capacity to 425 barrels per brew, and the full installation was complete by Easter 1954. With regret, Batemans had to cut down all the apple and pear trees in the brewery garden to avoid the risk of yeast contamination from a fruit fly infestation, and wild yeast on the fruit skins.

The improvements meant that the brewery was more efficient and productive, and Harry was keen to take advantage of the extra brewing capacity. Meanwhile, over at Greens – now calling themselves Flowers, adopting the more appealing name of yet another firm they had bought – the accountants had been looking at the dubious economics of supplying low-barrelage pubs in east Lincolnshire, far from their breweries and major roads. They concluded the best thing was to sell all but the best pubs to a local brewer for whatever price they could get. The irony, just a few years after grabbing so many pubs, was presumably lost on them, but it was borne of a fast-changing alcohol market, where drinkers were starting to turn their backs on pubs, preferring TV and bottled beer at home. Discussions between Flowers and Batemans led to the sale, in 1955, of twenty-three former Soulby pubs to Batemans for £23,000. They were not the best of outlets: only two sold more than a barrel (288 pints/163 litres) a week and nine of them failed to survive beyond the end of the decade. They also had a strange form of licence that prevented 'perpendicular drinking', meaning that patrons were required to be seated, as drinking standing up was rather oddly considered to encourage drunkenness. Harry sold all the closed pubs for a profit, and increased trade at almost all those that remained. The Red Lion at Mumby, which was selling no more than a couple of barrels per month, was given a thorough refurbishment, overseen by George's wife Pat – they'd married two years previously. Whilst the colour scheme might not sound appealing these days – lime green walls, maroon curtains and table tops, mushroom easy chairs – trade improved by over 400 per cent.

Father and son were not finished, however. Two years later, Flowers wanted to offload even more local pubs, selling a further sixty-six to Norwich-based Steward & Patteson, who'd bought Soames of Spalding. S&P were happy to sell Batemans twelve of those pubs, and also seventeen former Soames houses, the latter all pubs Harry

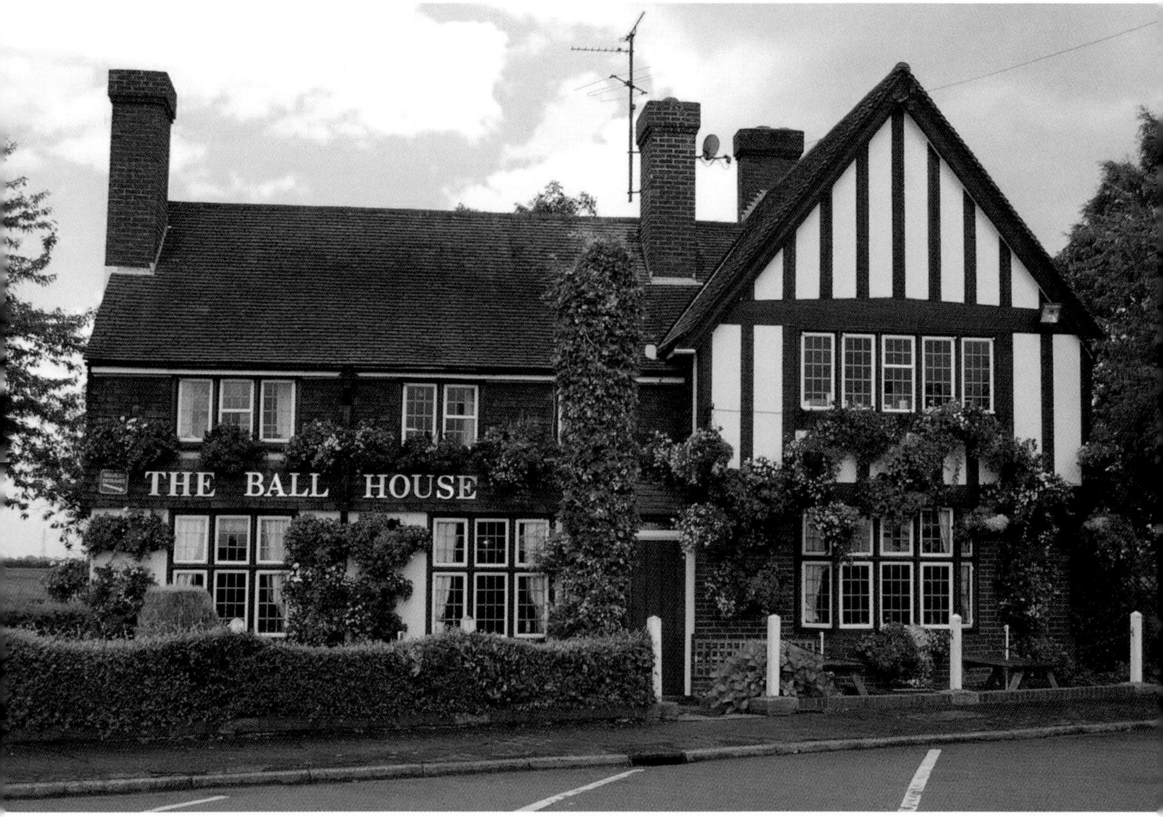

Proper Bargain
The Ball House, a mile north-east of Boston on the main road to Skegness, was one of the sixteen pubs bought from Steward & Patteson in 1957. It had been extensively rebuilt from a traditional cottage pub only three years before, as part of a road improvement scheme. The 'Brewers' Tudor' style had been all the rage pre-war, but by the 1950s cleaner, more modern designs were popular. (Batemans)

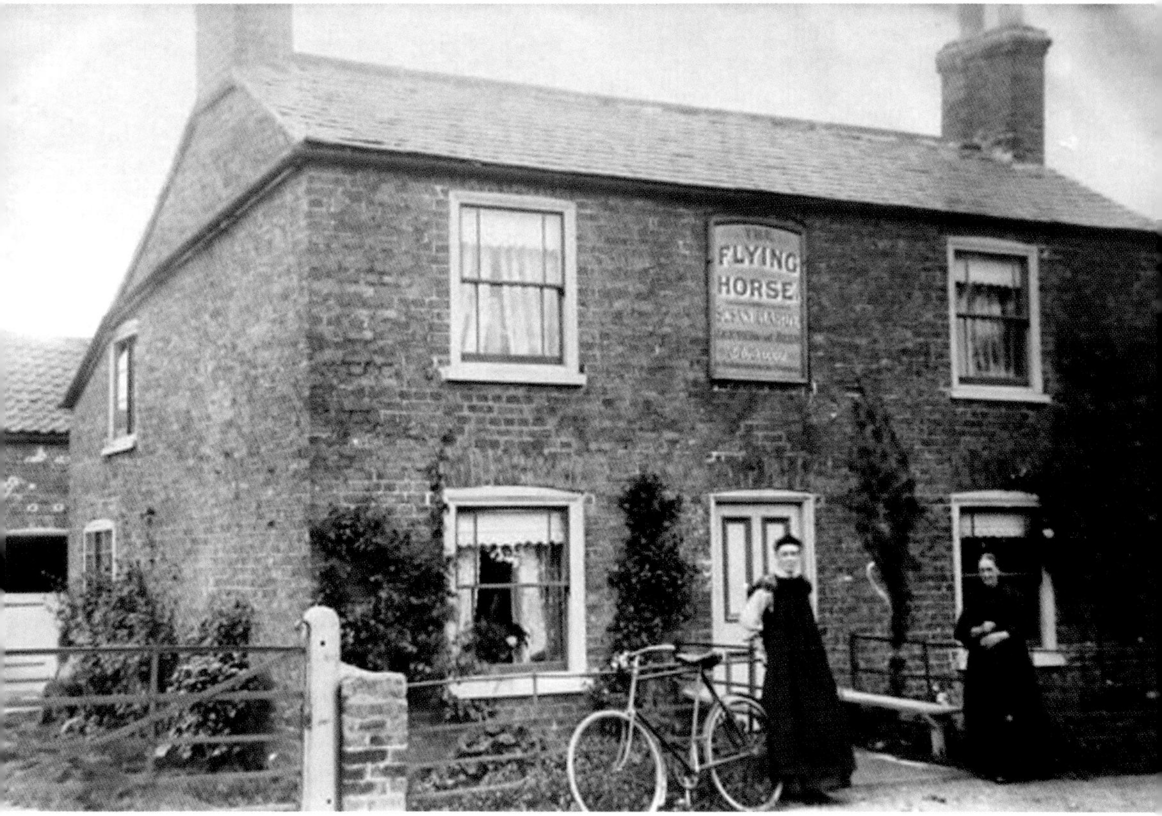

Other End of the Scale
The Flying Horse Inn, near Freiston, was also bought as part of the same deal with Steward & Patteson back in 1957. It was a small roadside pub that lasted three years with Batemans before closing in April 1960. Apart from the sign, nothing in this turn-of-the-century image identifies the property as a pub, and in the 1950s it sold less than a barrel a week despite its location on the main A52 road. After closure, it became a private house, which still stands today. (The Lost Pubs Project, closedpubs.co.uk)

knew very well from his time as a Soames director. It is quite likely he could not believe his luck. Three had been completely rebuilt in the last twenty years, five were in Boston, giving Batemans a significant presence there that has persisted to this day, and most were selling reasonably well; six mostly hopeless cases were sold. Some of the remainder, and some of the existing estate, were beerhouses licensed only for six days a week, so Batemans set about obtaining full licences where it was practical to do so; in some cases pub layouts had to be altered as the magistrates were reluctant to grant full licences to houses having just one room. Again, Batemans increased trade against a background of declining pub custom, and twelve of the former Soames pubs are still owned by the firm today. These developments all came at a price, of course, namely £43,900, which was part funded by a fifteen-year mortgage for £50,000 secured on forty-one pubs – the twenty-nine just bought and another twelve that until then had not been mortgaged.

Batemans ended 1957 with 108 pubs, the first time the magic three figures had been reached, although pub closures were starting to increase: over the next two years the estate receded to just under 100. This was offset to some extent by a still buoyant holiday trade in the coastal resorts, particularly in Skegness. For many years injured and recovering Derbyshire miners had been looked after in seaside hotels, paid for by mine owners who had to pay a levy to a national welfare fund, and in 1928 a very large specialised convalescent home for miners was built on a 9-acre site at Winthorpe, north of Skegness. In 1939 a holiday camp, the Derbyshire Miners' Welfare Holiday Centre, was added, catering for up to 900 people a week. At first, alcohol was banned there, a consequence of the religious principles of one of the founders, but when extended and improved facilities were added in 1951, beer made a welcome appearance. Harry ensured that Batemans supplied most of the camp's requirements, so that by the mid-1950s an average of 250 barrels a year were being sent there. This was a lot more than Batemans supplied to most of their pubs, but unfortunately it was a rare highlight as generally their free trade declined as the 1950s progressed; out of the forty or so pubs, clubs and off-licences supplied, three-quarters had declining volumes, perhaps most alarmingly at Butlins, where by 1960 barrelage was only 41 per cent of the 1955 level.

Some years later, George Bateman commented on the rural pubs of the past:

> There was a time at the beginning of the century when the pub offered greater comforts than the home and in the post war period it has been difficult to keep up. The majority of our houses had a public bar which consisted mainly of a scrubbed kitchen table and no counter, with outside toilets. All of them now have bathrooms, and the majority have central heating.

Of the twenty-six pubs closed by Batemans in the six years between 1955 and 1960, all but one were selling less than a barrel a week in their last year of trading. This was too few even for Batemans, who had done their best to help such houses, even to the extent of supplying their beer in unusually small 2-gallon casks they called 'piggins' – most casks at that time held 9 gallons. Three of them managed to survive as free houses (one of which, the Ebrington Arms at Kirkby on Bain, is still open – despite its Batemans branding, it's a free house). The great majority were small country affairs, at a time when such establishments still proliferated – it was still not uncommon to see a pub with just a few, if any, houses close by. The former patrons might have had to walk to the next nearest pub, but they had yet to be seriously inconvenienced. Batemans did their best to help their tenants, keeping rents as low as they could. In the 1950s, a third of their landlords were paying less than £20 a year; by comparison, of the fifty-two former Soulby and Soames pubs they bought, only four were in the same boat. Harry issued a prophetic warning to a local rating valuation court that many pubs would have to close, giving as an example the Plough at Halton Fenside, near Spilsby. It was sited at a crossroads a mile south of Halton village in sparsely populated but attractive fenland, a beerhouse licensed for six days a week with no drainage, electricity or mains water and having just two bedrooms in the apex of the roof: so restricted was the space that you could only stand up in the middle of them. The tenants, George and Fanny Rymer, were well into their seventies and paid just £18 rent annually; despite very low sales, they managed to struggle on until February 1960.

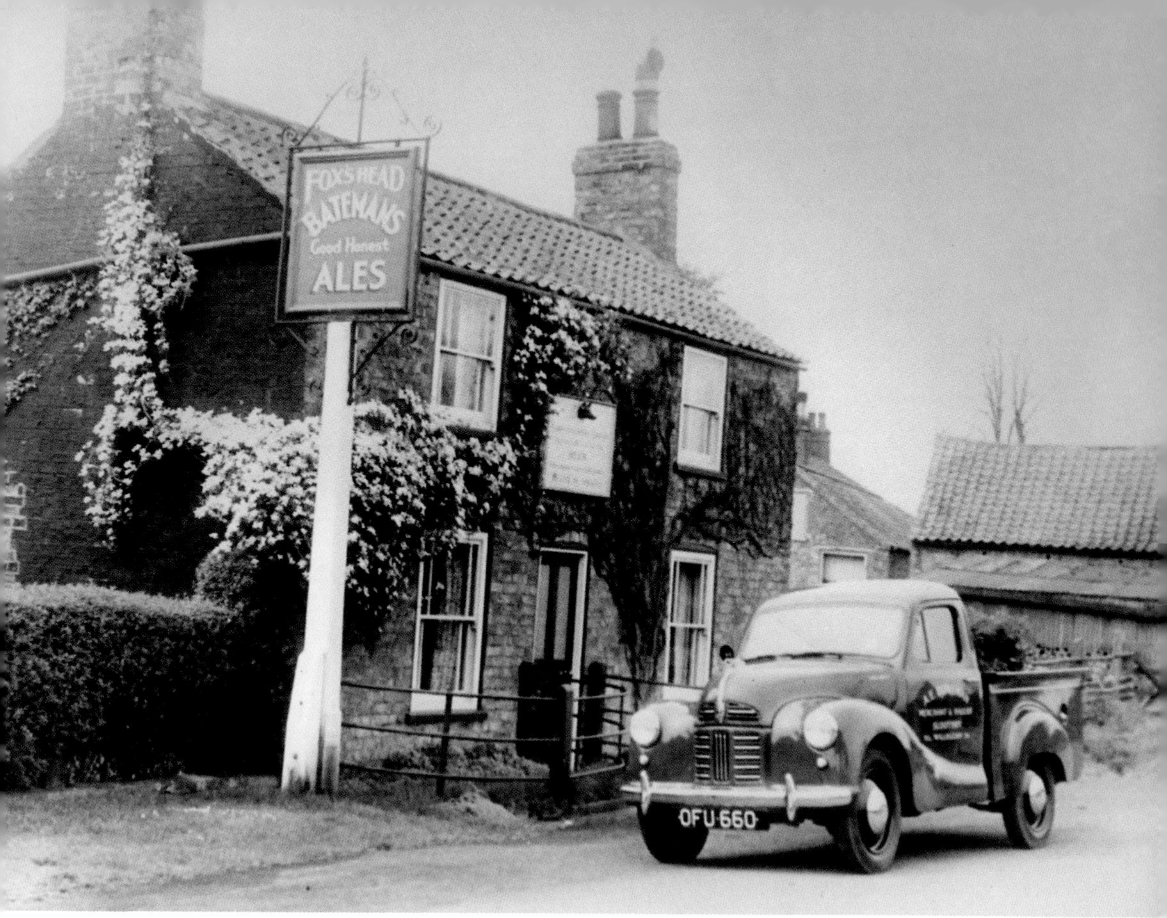

Slightly Foxed
The Fox's Head in the tiny hamlet of Sloothby, south of Alford, an idyllic country pub complete with its classic Batemans sign. But all isn't quite as it appears: although Batemans supplied a few barrels of beer every month, they didn't own the pub. The landlady, Louisa Jackson, spent an amazing fifty-eight years there until her death at the age of eighty-eight in 1960. The house then closed, passing to Albert Dowlman, a haulage contractor – it's his Austin van in front of the pub. Note the grisly addition of a fox's head on the front wall, hopefully a wooden or plaster moulding. (Batemans)

They were allowed to carry on living in the house for a while after it closed as a pub, Mrs Rymer dying there aged seventy-eight on the day it was finally sold.

The 1960s saw technical advances by the big brewers, who increasingly resorted to pressing on the drinking public huge quantities of lifeless and mostly tasteless keg beer, pasteurised at the brewery to stop any fermentation, and dispensed at the pub with added carbon dioxide to draw the beer up from the keg and make it sparkle. Though beer served in this way was bland and lacking in any character, it was at least consistent – the chances of a bad pint were much reduced. Customers were also getting a taste for lager, typically a licensed copy of European, Canadian or Australian brands, or – as the 1970s dawned – mostly invented German-sounding names. These were all beers that could be manufactured relatively easily, in industrial quantities, often in huge new factories, with

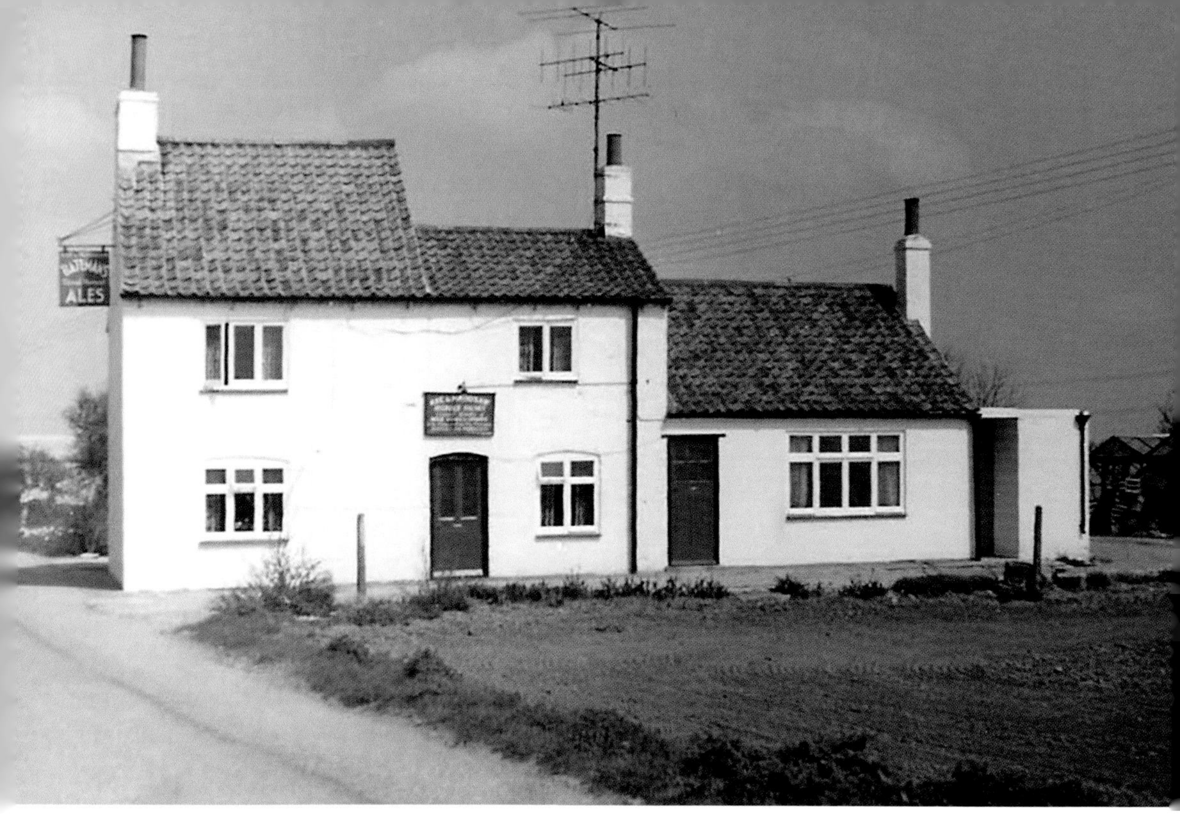

Keep it in the Family
Many families in Lincolnshire can claim a connection to a Batemans pub, and this is mine. The Axe & Handsaw at Quadring Eaudike was bought by Batemans in 1951 when my late uncle's father, Reg Palmer, became the tenant. It was a typical fenland pub, as this image shows. Untroubled by other housing, it sat by the roadside at the corner of a field. It had at least three stages of development: the original house to the left was extended to the right in two stages, with a toilet block added later. The enormous TV aerial is securely strapped to the chimney stack. The pub was run by the family until 1972, when Batemans sold it at a small loss. Despite the seemingly unpromising location, it remained open, renamed the Goose Fayre, until the late 1990s. It's now a private house. (The late Barrie Palmer)

minimal labour at significant profit. The burgeoning sales of keg meant it was not a good time to be a traditional brewer, most of whom were unable to compete and invest in advanced keg beer plants, although they did try low-tech solutions. Batemans constructed a small and basic keg washer and filler, as well as a second-hand milk pasteuriser they'd bought from a local dairy farmer. For a while, some kegs were pasteurised by submerging them in boiling water, but it was soon decided that there was a risk of explosion, and that particular practice was quickly ended. The makeshift set-up was of some help to the lowest low-barrelage pubs, as kegged beer should last longer than cask real ale and be easier for landlords to look after. But the plant never worked very well, causing issues with quality and shelf life, and it was quietly phased out.

Smoking was still a majority pastime for British men in the 1960s: more than seven in ten were smokers, as well as four in ten women. With health concerns largely unrecognised, cigarettes were widely available in pubs and shops, even to children as young as sixteen. Gradual increases in tobacco duty had created a ready black market for illicit sales, with high-volume thefts becoming more widespread. Organised criminals targeted Batemans in August 1962, forcing the doors of the brewery's wine stores to steal 600,000 cigarettes. The gang even used one of Batemans' vans as their getaway vehicle. It was George Bateman who discovered the theft just after 6 a.m. that morning, over two hours after he'd started brewing. The criminals had struck between 10.45 p.m. the night before – when George went home – and around 4 a.m., managing to carry out the raid without being heard by any nearby residents. The missing stock, a selection of Players, Senior Service and Park Drive, for those who might remember brands of that time, was never recovered, although the van was later found dumped at Rauceby, near Sleaford, no doubt where a switch of vehicles was carried out. It was a well-planned raid, taking advantage of the high stocks Batemans needed for supplying its pubs at the height of the summer holidays; the wines and spirits in the storeroom were all left undisturbed. Despite police reporting 'promising lines of inquiry', nobody seems to have been arrested in connection with the offence, and further large-scale thefts continued across the East Midlands that year, although Batemans were not singled out again.

The number of pubs continued to decline in the 1960s with the closure of a further thirteen houses, again mostly country premises, but as a sign of the times, also serving its last pint was the Peacock and Royal Hotel in Boston marketplace. Patronage at the rambling old thirty-room hostelry had fallen away steadily since the war, the main part of the hotel closing on New Year's Eve 1962, with just the hotel tap and Adam Bar staying open for a few more months. The premises were sold in 1965 for £64,000, ironically at an auction in the Rum Puncheon pub on the other side of the marketplace, and later demolished, to be replaced by a large Boots store. On the brighter side, Batemans continued to pick up decent pubs when they became available, including five from national brewer Whitbread in 1970, having some years earlier declined to join their 'umbrella'. This was a rather controversial creation whereby Whitbread would agree to buy a significant minority of shares in a smaller brewer, to 'protect' them from takeover. However, what tended to happen within a few years was a complete takeover by Whitbread, leading many independents to view the umbrella with deep suspicion. Fortunately, Batemans were never tempted.

In 1972 Batemans spent £53,750 – easily the most they had ever paid for anything – on buying the Parade Hotel, Skegness. The vendors, the Salt family, had invested significantly in the forty-seven-room seafront hotel, adding a large ballroom and restaurant. Meanwhile, trading at the Boston wine merchants Ridlingtons was holding up. Their best-sellers were still spirits and fortified wines marketed under the 'Pilgrim' label, devised by Harry Bateman in 1921, which from the 1950s featured a Puritan figure with the Stump in the background – the association was not inaccurate, as there's no evidence that religious groups such as the Pilgrim Fathers were teetotallers. The business had celebrated 250 years trading in 1955, albeit under several different changes of ownership, including staging Boston's first wine tasting event, to the fascination of the local newspaper – not until the late 1970s would most Britons show an interest in anything other than sweet German wines and cheap French reds. Harry's younger son John had been in charge there since 1950. He'd also inherited his father's entrepreneurial spirit, having both an understanding of wine production and the spirits market. In 1978 Batemans bought a Grantham-based wine merchants, Frederick Fletcher Ltd, and expanded the Boston business, developing the site to the rear of the South Street premises as a wine warehouse open to the public. In the early 1980s there was a proposal to split the brewery and Ridlingtons into separate companies, but nothing seems to have come of it.

Batemans had celebrated their centenary in 1974 in fine style: a lunch for 200 followed by an evening jamboree attended by 800 staff, landlords and other customers in the brewery gardens. A photograph taken of the staff, with Mr George and Mr John, as Harry's sons had become known, at the centre, holding a Batemans pub sign, and the reassuring presence of the old mill in the background, seemed to strike exactly the right note. When asked if they would be brewing a special beer to mark the occasion, Batemans claimed there was no need, as their existing beers were so good! Colin Crowson, who had recently left Watney Mann, had been hired on a year's contract to manage the celebrations, but he proved so effective that he stayed for another twenty years, becoming a director of the company in 1989.

Business was holding steady at ninety-nine pubs and three hotels, but sadly Harry Bateman did not live to see the landmark event. His death in May 1970, at the ripe old age

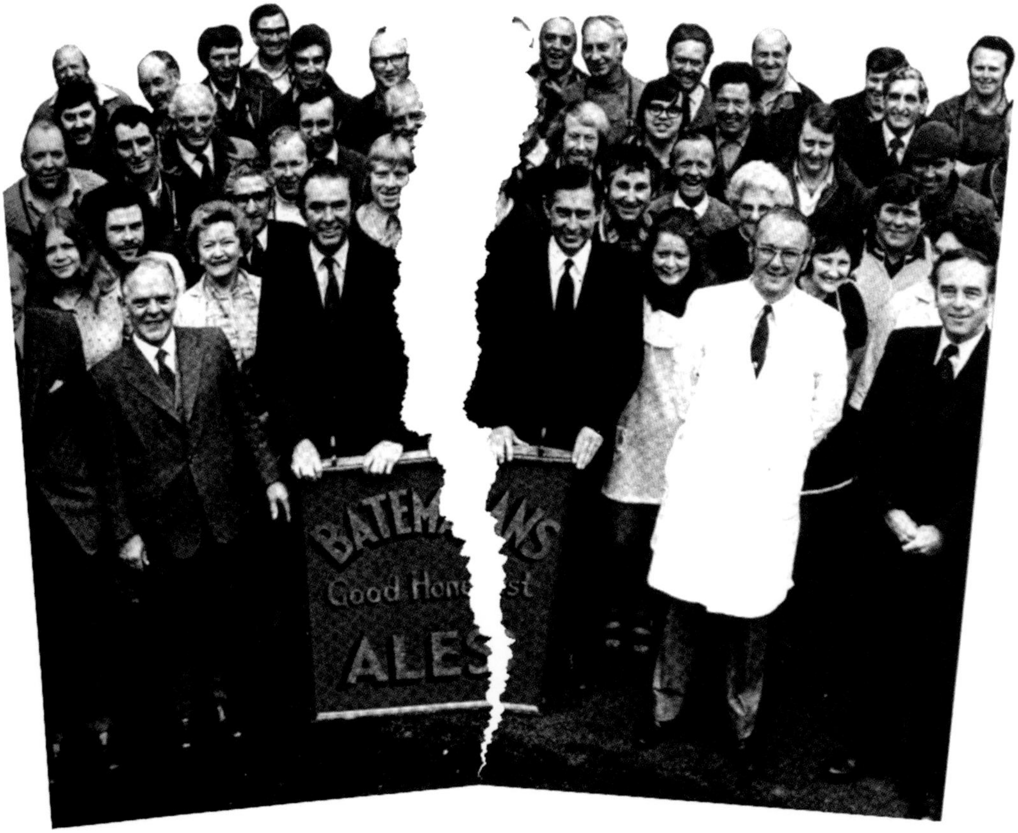

The Split
This graphic image, a doctored version of a photograph taken during the centenary celebrations in 1974, emphasises the family disagreement about the brewery's future that flared up in the late 1980s. John Bateman is on the right, George on the left. (Batemans)

of eighty-six, had robbed not only Batemans but the local community of a central figure and leader. A widower for the last six years of his life, he had gradually stepped back from his many local roles, but was still chairman and managing director of the brewery when he died (although George was actually running the brewery at that time). In his will, he left his estate – including his shares in George Bateman & Son Ltd, in which he had a controlling interest – to his three children. The shares were split so that George, as Managing Director, had slightly over 40 per cent, John slightly under, and Helen had the remaining 20 per cent in trust. She took no part in the business, having married Harry Caudwell, a local farmer and former solicitor. Because the family held all the ordinary shares – which came with voting rights – they could control company decisions; preference shareholders, of which there were many, got a better dividend rate but had no say in how the company was run. Batemans was effectively a private company, safe (or so it seemed) from takeover bids, which was fine as long as everybody was in agreement about company policy and strategy. After the death in 1979 of Jessie Bateman, Harry's sister, her shares were divided equally between George and John, which continued to respect the requirement of Harry's will. Unfortunately, those arrangements contributed significantly to a major problem fifteen years later that was to take the brewery to the very brink of closure.

The immediate problem for Batemans in the 1970s was the continued gloomy outlook for smaller brewers, under pressure on all sides by lack of investment capital, changes in drinkers' preferences and the dominance of national brewers. Some even abandoned cask ale completely or stopped brewing to continue as a pub-owning company, taking their beer from elsewhere (a course adopted in 1974 by Melbourn's of Stamford, in the south of the county). When George was elected chairman of the East Midlands Brewers Association in 1970 – a role his father had stepped down from seventeen years previously – there were far fewer brewers to represent, as so many of them had sold out to big business.

Then, something unexpected started to happen: slowly, people were starting to seek out traditional cask beer again. Pressure groups like the Campaign for Real Ale (CAMRA), founded in 1971, and a growing band of writers championed traditional pubs and real ale. CAMRA membership spiralled quickly to over 20,000 in the space of a few years, fuelled by a dislike of national brews rampaging across the countryside – in cities such as Norwich, for example, it had become very difficult to find any real ale, and what was available was quite bland. The fact that Batemans' range of beers had changed little over time, whilst remaining distinctive and flavoursome, was part of the attraction. They offered three draught ales:

- Light Mild
- Dark Mild (DM)
- XB Bitter

and six bottled beers:

- IPA
- Strong (a brown ale)
- Light Dinner Ale
- Nut Brown

- Double Brown
- BBB (a strong barley wine, only available in 'nips' – ⅓ of a pint)

Independent brewers such as Batemans found their beers were increasingly sought by the limited free trade that then existed. When, in 1976, a pub in Cambridge, the Salisbury Arms, started selling Batemans XB, George and his wife Pat excitedly arrived for some quality control, but quickly found the beer they were offered certainly wasn't theirs. George demanded a cellar inspection, where he found the pipe to the hand pump was connected to a barrel of Sam Smith's bitter instead. 'They didn't make the same mistake twice,' he explained.

Batemans met the renewed demand for cask beer by devising their first new brew for many years: XXXB, launched in 1977, was a full-flavoured strong bitter, brewed at 1048deg Original Gravity, similar in strength to other premium ales that independents were launching at the time. Devised by head brewer Ken Dixon, who'd just taken over from the long-serving Dennis Hartley, it was available (and still is) on hand pump and in bottled form. It was immediately popular within the relatively small market of Batemans' pubs and free trade. The growing number of beer festivals was therefore very important for Batemans as it enabled them to reach much wider markets; they sent nearly 500 gallons (4,000 pints) of XXXB, XB and Dark Mild to the 1979 CAMRA Great British Beer Festival in London, for example – double the amount supplied in 1977.

Mr George was very aware of Batemans' role in supporting communities and keeping pubs open, but there was a difficult balance to be struck, as he explained in 1981:

> We now own all but one of the pubs on the road from Wainfleet to Boston. There are six villages each with a population of not more than 500 people and each with its own pub. We very sincerely wish that each village shall continue to have its own pub but there is no way at all that they can continue to be viable serving the local community only. The only way to keep each pub viable is to try and give different emphasis to each.

Unfortunately, even that approach proved impossible to attain, as a past head brewer, Martin Cullimore, explains:

> George made it clear they would try very hard to not close the last pub in the village. But some of the pubs were really not viable by this time. My job as Head Brewer included going round the houses to make sure the beer they were selling was of the right quality, and I remember going to see one landlady – she didn't open her pub during the day so I had to visit her in the evening. She told me she took the tenancy because the rent for the pub was less than she would pay for a private house.

Some marginal pubs were therefore sold with their licences, to continue if they could as a free house. In the 1980s Batemans had nearly 500 applications a year from prospective tenants (it's now much lower at around fifty) and it was very difficult to match tenants to pubs. George Bateman said he sometimes just used his intuition, but he liked them to have had some prior training, if not experience. Many might be going into business for

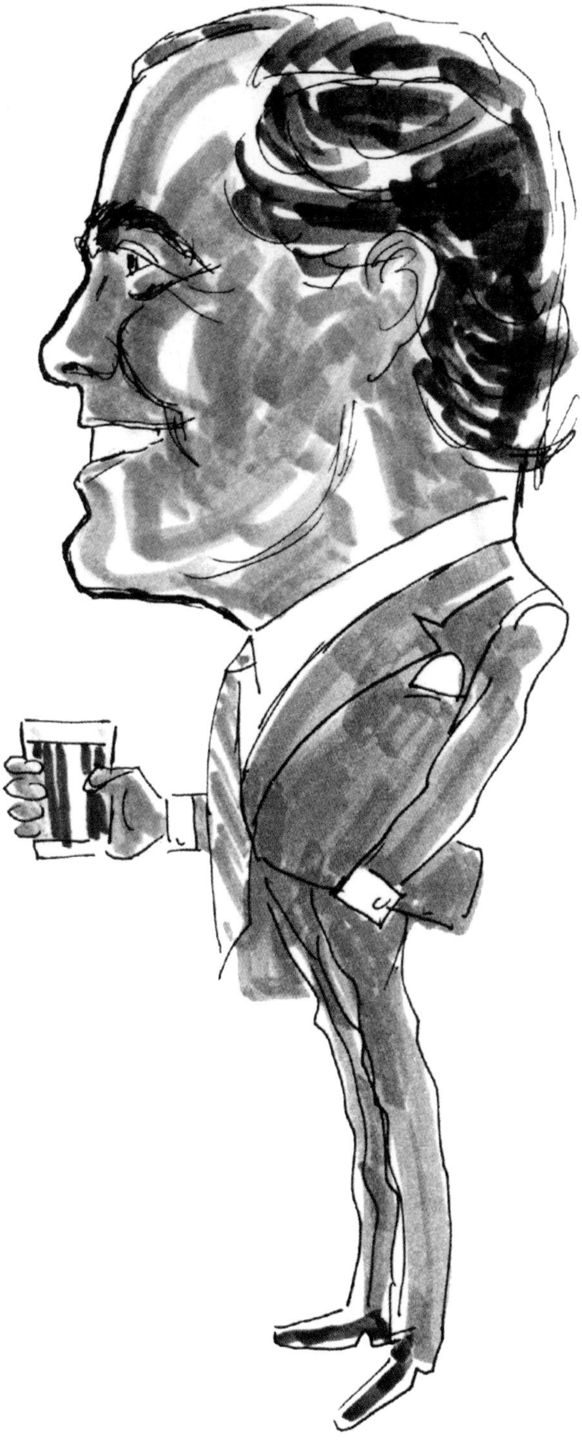

The Man Himself
An appealing sketch of George Bateman, holding a symbolic pint to confirm the likeness. As always, he's smartly dressed. (Batemans)

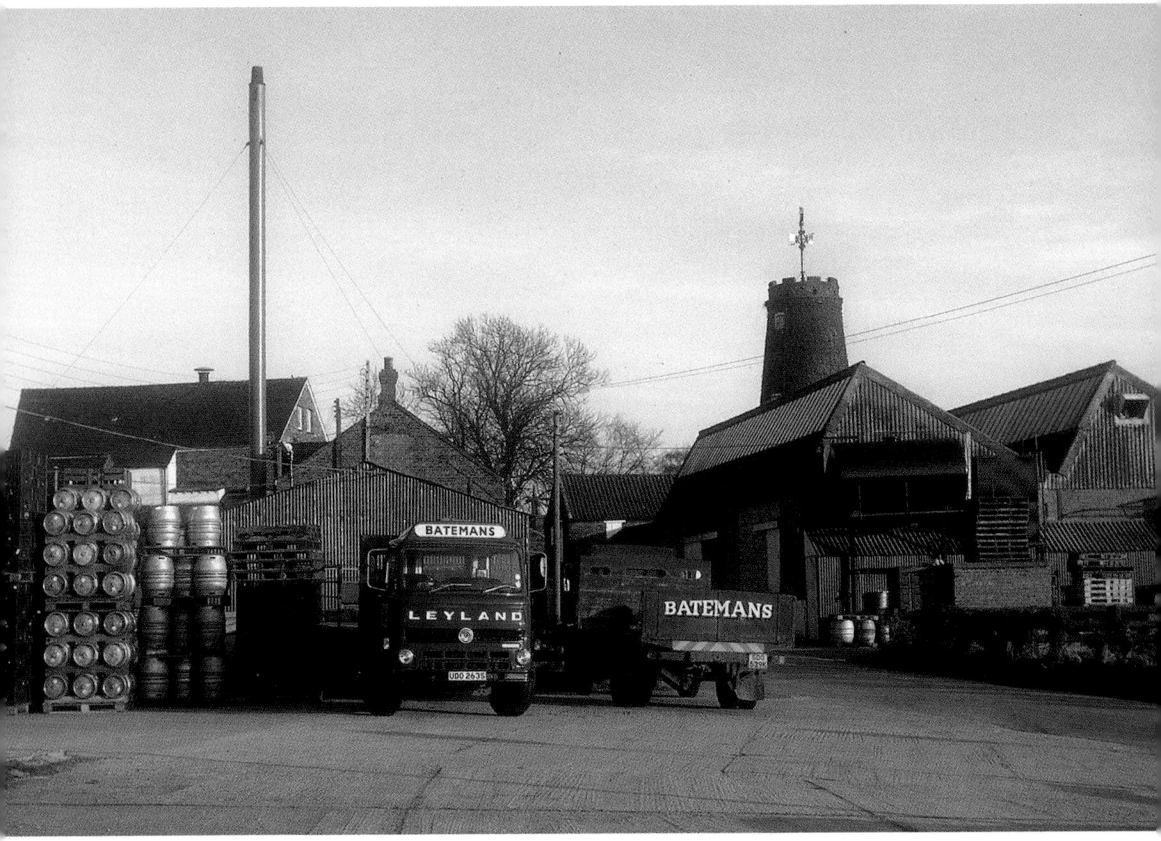

Cold Refreshment
The brewery yard on a bright winter's day in 1980, with a couple of Leyland drays awaiting their next loads – probably the barrels on their left. Mill Lane runs through the site to the right of the lorries. In front of the mill is the twin roofs of the bottling store. The brewery is just behind the drays. (John Law)

the first time, so they were warned that they were not taking on 'a little goldmine' and the hours would be long. Most pubs were run by husband and wife at the time, and most tenants tended to run out of steam after ten or twelve years. Houses no longer remained with families for generations, and Batemans did not have enough houses to guarantee a move to pastures new.

Mr George was delighted when his two children decided to join the family business. His daughter Jaclyn – who had trained and qualified as a nurse – decided to join the business in 1983. She became Batemans' first ever credit controller, a role previously undertaken on a fairly ad hoc basis, but she was soon spending more and more time developing new markets for Batemans, especially in London and the south where the firm had previously been almost completely unknown. A year later, his son Stuart started his training as a brewer, like his father before him, although he'd been informally trained since the age of six, when he spent a day helping Peter Sharpe, the brewery's maintenance engineer.

'Does he swear?' Peter had asked George.
'Of course not!' was the surprised reply.
'Well, he ******* well will be by the end of the day!'

This gradual induction continued during Stuart's school holidays, when he was put to work in every brewing department from bottling to brewing, cask washing and warehousing, later obtaining his HGV 1 licence so he could drive the brewery lorries: the ability of family members to perform all these tasks could be crucial for a relatively small enterprise. After studying business and hospitality at St Paul's College, Cheltenham, Stuart had spent a couple of years as a sports teacher – he says he wanted to do something completely different for a few years before going into the family business. He'd been earmarked for training at Whitbread, who were then a huge national brewing business, but it was decided that a regional brewer might offer a better, more relevant experience. He therefore spent three years being schooled in all aspects of the brewer's craft at Mansfield Brewery, 85 miles away in Nottinghamshire, where George had a long-standing friendship with the

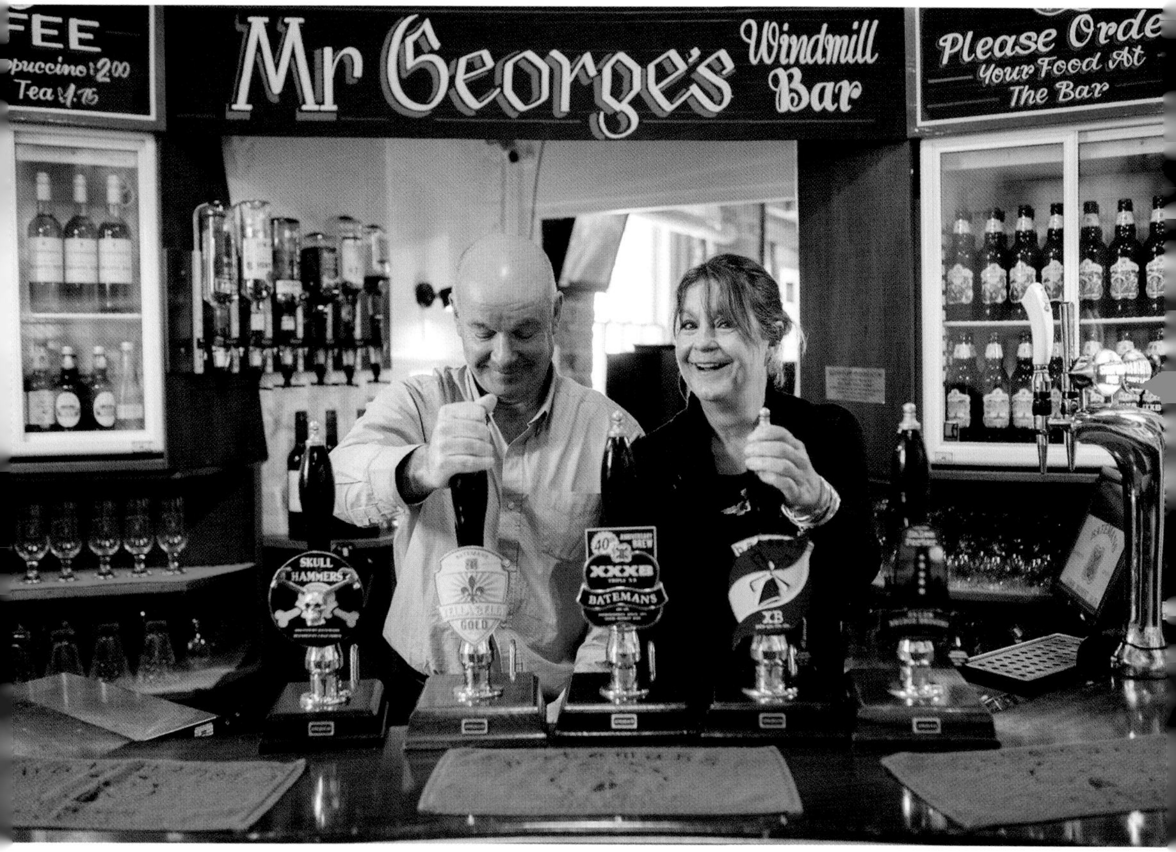

Double Act
Over at Mr George's Windmill Bar, on the ground floor of the old mill, Stuart pours a pint of Yella Belly Gold while Jaclyn stands ready at the XB hand pump. Skull & Hammers, XXXB and Salem Porter were the other beers on offer that day. (Batemans)

chairman, Robin Chadburn. Their modern brewery, serving an estate of over 400 outlets across the East Midlands and Yorkshire, was a far cry from the historic but cramped Wainfleet site, but they had just reintroduced real ale after a ten-year keg-only hiatus, and Mansfield had much more in common with Batemans than the giant Whitbread concern. It wasn't just brewing that Mansfield taught Stuart; they went through all elements of the business, covering marketing, financing, free trade, managed houses and tenancies.

Further good news came with the arrival of a new head brewer in 1985: the previous postholder, Ken Dixon, had left some months previously. In his stead came Martin Cullimore from Mansfield Brewery, where he'd met Stuart Bateman. The two men were roughly the same age, Martin being a few years older at thirty, and they got on well. Stuart told him of the vacancy back at Wainfleet, and Martin picked up the phone to George Bateman. That was the starting point of a remarkable thirty-two-year tenure, at the end of which he was the longest-serving head brewer in the whole country; such longevity is rare in the industry. One of Martin's first tasks was reviewing and refining the brews. He remembers George being very supportive –

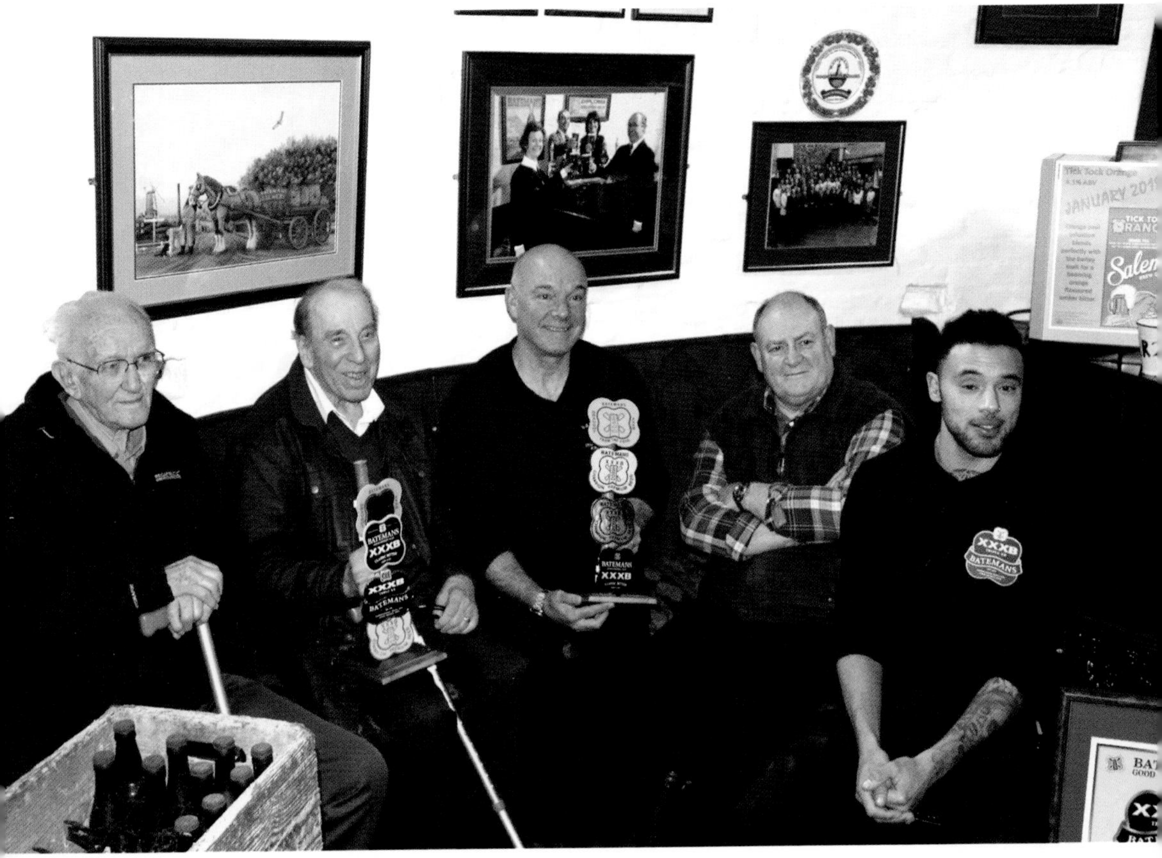

Years of Experience
To celebrate XXXB's fortieth birthday 'upgrade' in 2018, when the premium ale returned to its original 4.8% ABV, head brewer Scott Lawrence (far right in this picture) invited back his three predecessors Dennis Hartley (first left), Ken Dixon (second left) and Martin Cullimore (fourth left). Stuart Bateman is third from the left. (Batemans)

after all, he had trained as a brewer and still had a real interest, being in and out of the brewery much of the time – and the two men agreed a plan. Out were ingredients like flaked maize, which Martin didn't see as right for a traditional brewer like Batemans (though ironically the first George Bateman had been using flaked maize and rice back in the early 1900s). He recalls: 'It took me 18 months to change those recipes – you can't make sweeping changes all at once as drinkers will notice changes in how the beer tastes. But when I'd finished, the beer was made using only malted barley and hops. I thought that was important for Batemans.'

It was as a result of attending a trade fair in Milan in 1984 that an order was won to supply beer to Italy, branded by Batemans as 'English lager', which they had managed to brew using an ale yeast strain. The initial requirement was for 23,000 bottles and Martin Cullimore remembers that very early in his career George asked him to fly out to Milan in connection with the contract. The brewing equipment in his luggage set off the airport's security scanners in dramatic fashion and before Martin knew what was happening he had the Italian border police pointing guns at him. A small range of bottled bitter followed and a fearsome threesome of kegged Super Bitter XXXX, Double Lager and Strong Brown XXXXX. The latter at 8.6 ABV (alcohol by volume) was so strong that after being pitched into fermenting vessels on a Friday, the entire fermenting room was found to be full of froth the next morning! Further drama followed when George fell over on Salem Bridge, knocking several teeth out just before he was due to drive the first batch of 80 gallons over to Italy. Despite the flow of blood, George was determined to carry on, so accompanied by his son Stuart and waved off by all the staff, he set off in a brewery van for Milan. The 900-mile trip, driving in turn, took them an arduous thirty hours.

Sadly, although sales were good, the logistical problems of administering a contract for delivery so far away, and the importer's failure to return the empty barrels, eventually proved

Italian Job
The first commercial collection of beer leaves the brewery in an Italian lorry for export, following George and Stuart Bateman's journey to Milan. (Batemans)

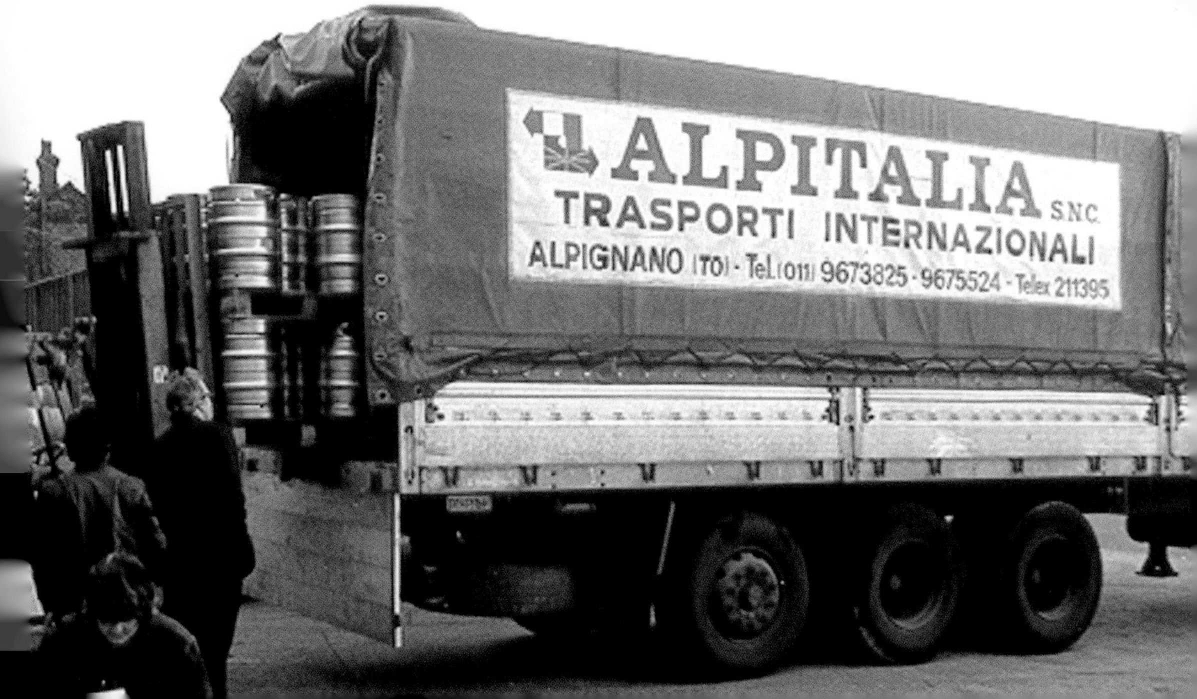

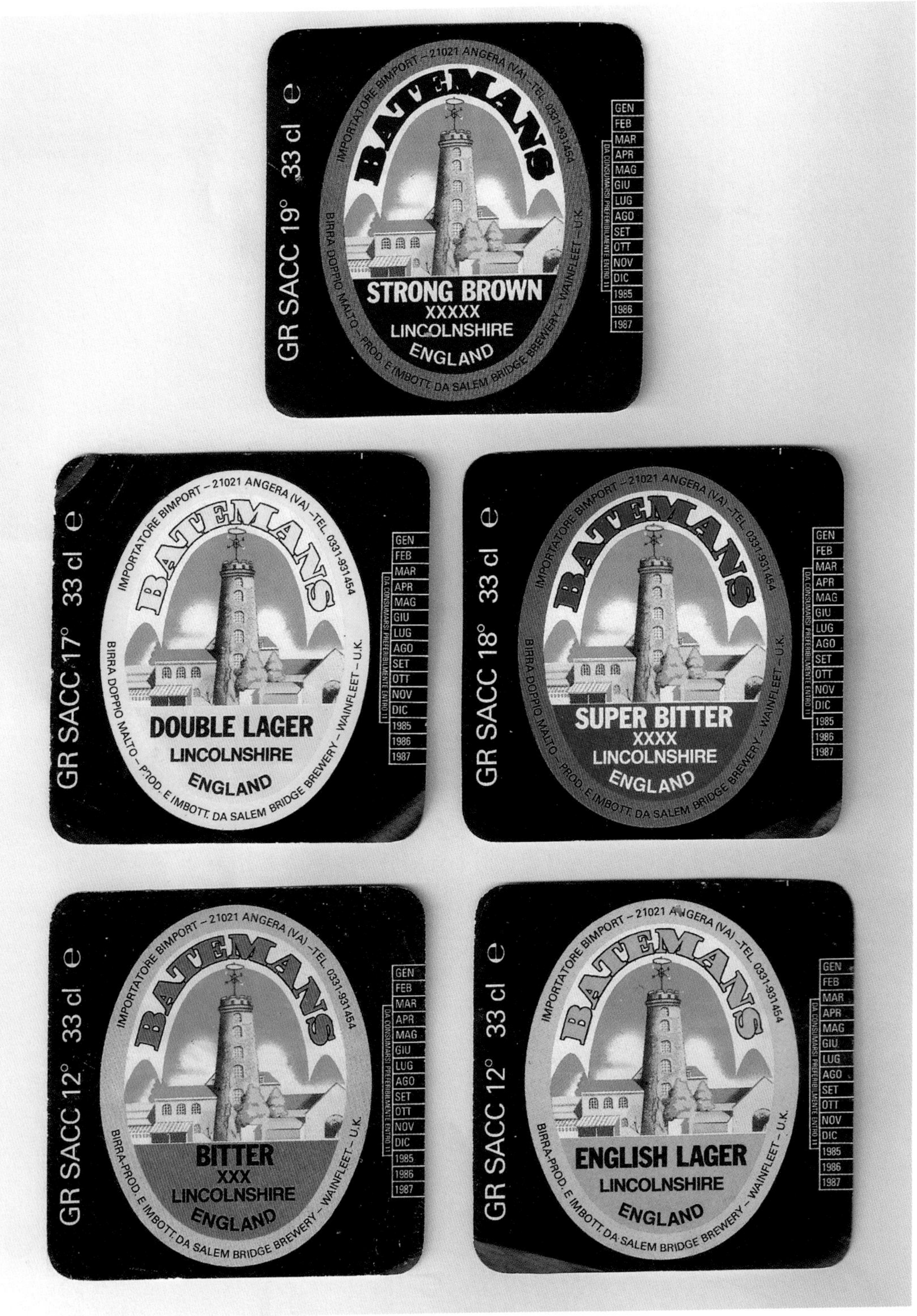

Birra
Bottle labels for an Italian beer contract of the 1980s.

terminal. But at least Batemans had reached new markets and the free trade was starting to expand. Their first pub in Lincoln – the Golden Eagle, in the lower High Street – had opened in 1979 and funding was also found to acquire another five pubs from Whitbread in 1984 and pay for them over a seven-year, interest-free period. In return, Batemans agreed to buy 4,000 barrels of Heineken lager per year for the next twenty years. A new, brighter corporate livery of scarlet red with cream letting started to appear on pub signs, lorries and vans, but retaining the 'wavy logo' featuring the four-sailed windmill that had been a feature of Batemans for many years.

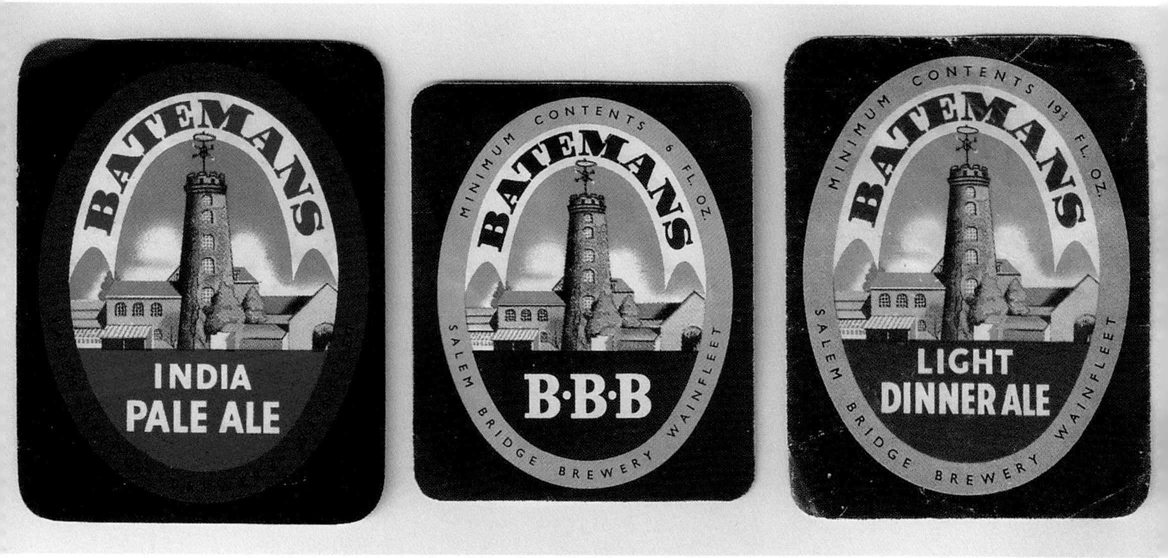

Beer Labels of the 1960s (above) and 1980s (right)
A comparison with the images on p. 29 shows the 1960s labels are updated versions of earlier styles, apart from the discontinued Old Mill, taking advantage of advances in colour printing. By the 1980s, however, a simpler style had evolved (right) using the windmill-themed 'wavy logo' as a reference point across the brands. Note the appearance of 'best before' dates, where the label would be nipped with the appropriate month and year, as shown with the Ploughman's Ale. Metric bottle sizes had replaced imperial; three of the four labels are for smaller 275 ml bottles, which are now much less common, the exception here being Nut Brown. None of these beers are produced today.

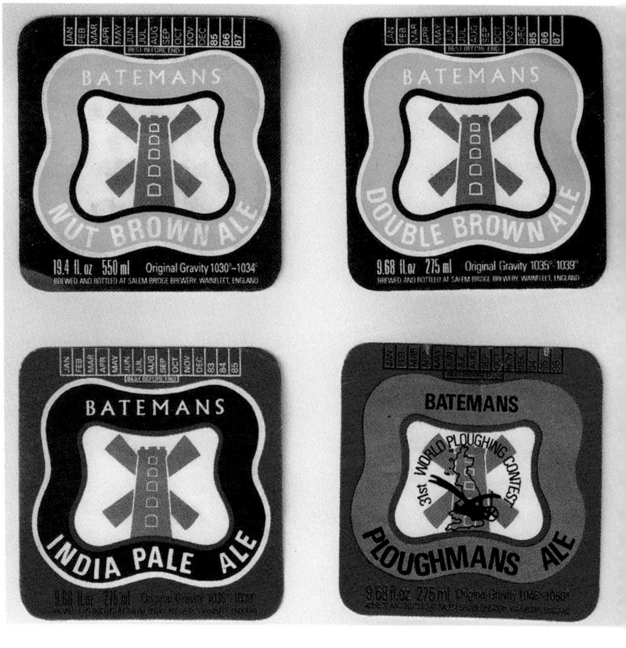

7

Independence ... at a Price, 1985–2000

The omens for the future must have felt good, but everything changed in March 1985 when George was contacted by John Bateman and their sister Helen Caudwell; they wanted a meeting. The three siblings gathered at the brewery. It quickly became clear that John and Helen wanted to sell their stakes in the business and retire. Between them, and John's son Peter, the owner of a food service business in Boston, they controlled nearly 60 per cent of the ordinary shares and were therefore able to outvote George. They may well have made the point that with nearly 100 pubs, three large hotels and a respected wine and spirits trade, the business could be worth several million pounds to the right buyer. When independent breweries had sold out in the recent past, the owners had been able to enjoy a comfortable retirement. None of them were getting any younger; Helen, the eldest, had already turned sixty and the two brothers were in their late fifties.

Although not completely unexpected, it was nonetheless a shock for George, who remained passionately attached to the brewery. Whilst he understood the logic of selling, he also knew that the first act of any buyer would be to shut down the brewery, resulting in the loss of sixty jobs. No corporate boardroom would look kindly on the future of a small, rather cramped brewery, needing investment, somewhere up on the remote east coast. George simply felt he had a responsibility to his staff, landlords and the local community to keep going if possible. No doubt he had been tracking proceedings that year up at North Country Breweries in Hull, just bought by Mansfield Brewery, who promptly announced its immediate closure. He was determined that would not happen at Wainfleet, but it seemed that the only course available was to buy out John and Helen's stakes. To do that he would need to find an enormous sum of money somehow.

It had been agreed that George could have a few months whilst he investigated possible routes to financing, but the more people he spoke to, the greater became the risk of a leak, so that July he gathered staff and tenants together to tell them of the devastating news. Martin Cullimore had been with Batemans for just six weeks and remembers George asking if he wanted to leave for a more stable job; he told him he wanted to stay. Len Osborn, landlord of the Britannia in Boston, fifteen years a Batemans tenant, spoke for many publicans: 'If we have any problems, we can always ring up Mr George. We don't have any junior assistant managers wittering away at us. Batemans are the only brewery left in Lincolnshire and it would be an absolute tragedy if we lost it.'

The local media pressed George for some quotes. He expressed his fears frankly: there was a very real danger the brewery and all its pubs could be sold. But, he added, 'if we have a chance of saving the business, we will look for new partners and shareholders. If I can raise the money, and service the debt, I'll do it.' When national broadsheets came calling, George set out his philosophy, telling *The Guardian*: 'There's something a bit special here. It's not all about the bottom line. I still think the building of a happy community is something worthwhile. I hold this business on trust for future generations and the local community.'

The whole episode was extremely taxing on the whole of the Bateman family; George, Pat, Jaclyn and Stuart. George, supported by Pat, travelled up and down between Wainfleet and London as he spoke with City investors and banks. Jaclyn set about establishing a network of wholesalers to boost sales of Batemans' beers, whilst George and Stuart arranged meetings with other breweries and pub companies which they thought might be interested in working with them whilst, most importantly, keeping the brewery open. However, most possible 'partners' were more interested in asset stripping. George and Stuart met with Control Securities, Belhaven Brewery, Mansfield Brewery and Whitbread, and further down the line there were discussions with a wholesaler called Liquid Assets. But all of the discussions and negotiations were fruitless as guarantees of continuity could not be given. The representatives from one company were physically thrown out of George's office when they informed him that in the event of them taking over Batemans, he would be kept on to do a bit of PR for them! In such meetings George always tried to live up to his favourite expression, 'Death or Glory' – the motto he'd adopted from the 17th/21st Lancers, the regiment from his National Service days. He also wore his regimental tie to all meetings.

The nearest they came to a white knight was Belhaven, which was at least a traditional brewery albeit one expanding fast under the charismatic new direction of hotel entrepreneur Nazmu Virani; they needed extra capacity and had spoken publicly of retaining brewing at Wainfleet. Stuart remembers an unconventional breakfast meeting at Virani's impressive offices in London, with the proprietor jogging away on his running machine throughout. Midsummer were a less-favoured alternative as one of the first pub-only chains (pubcos), a CAMRA spinoff that had been sold into private ownership; they would be unlikely to keep the brewery going. Mansfield Brewery was well known to the Batemans, but negotiations made little headway owing to their insistence on having a majority on a restructured Batemans board, a prospect George found unacceptable. And the aspirations of Whitbread did not fit in at all with how the Bateman family saw the future for the brewery, staff and pubs.

The pressure was unrelenting, and too much to bear at times; people would stop George in the street and express their total confidence in him, which just made the burden worse, he recalled. George's wife Pat became so upset at the toll it was taking on her husband that she took out her frustrations on an oil painting of Harry Bateman that hung in the brewery, slashing it and throwing it in the Steeping River; fortunately, George managed to wade in and retrieve it. As negotiations dragged on with no resolution in sight, George even considered selling his stake. But the Batemans were cheered by the best wishes of both his staff and the local community, many of whom were appalled at the prospect of the firm closing down but could not know the vast sums needed to save it. There was an elderly farmer who offered to invest £3,000, declaring 'George, your troubles are over'; and the two

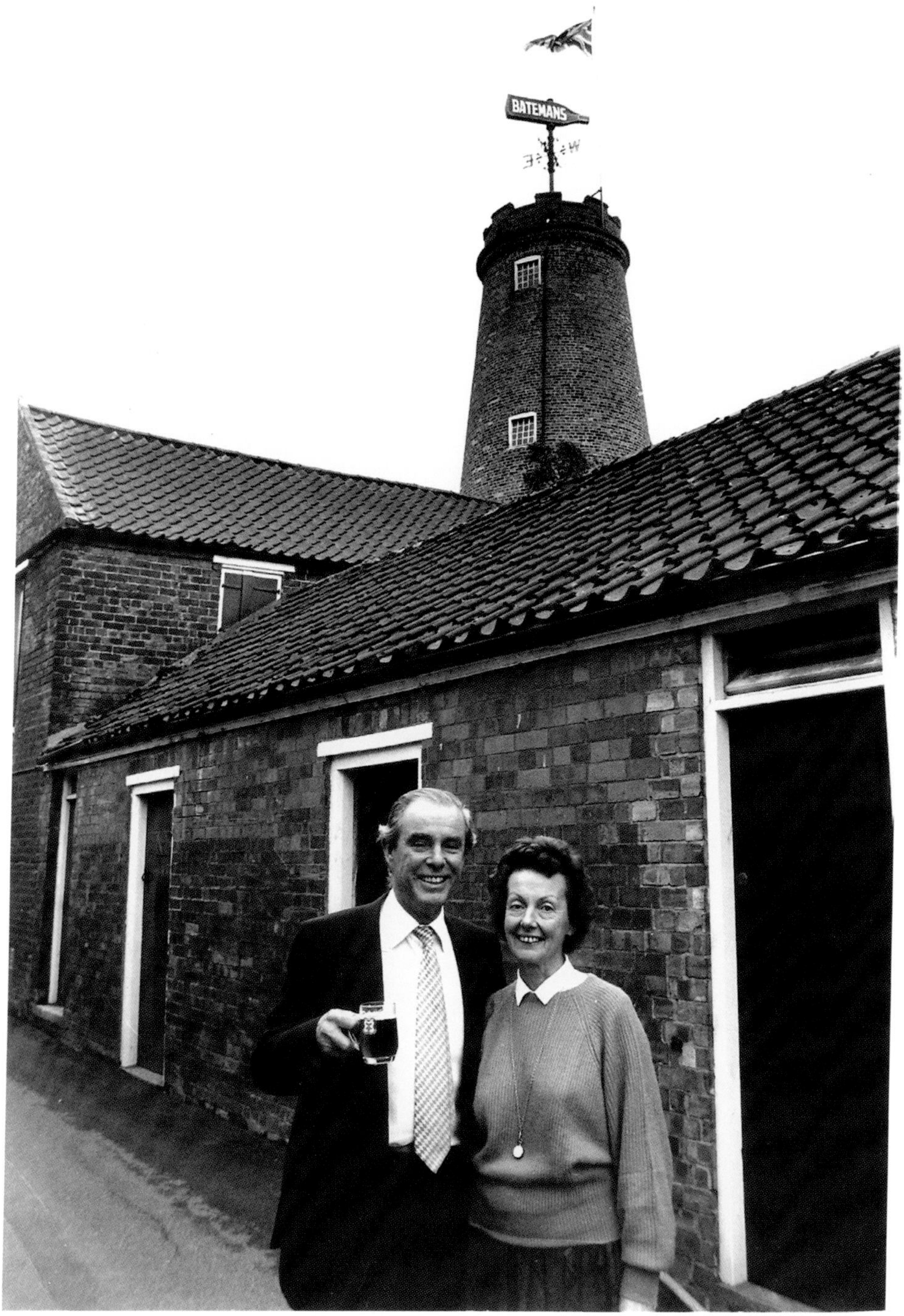

Proud Couple
A nice image of George and Pat Bateman in Mill Lane, taken shortly after the firm had won its independence in 1987. George holds a celebratory pint, no doubt at the photographer's request. Behind them stands the mill, with the ivy well under control, allowing us to see the castellated mill top and the beer bottle weathervane. (Batemans)

daughters of a tenant, who offered to put a bottle on the pub's bar to collect money. They were particularly touched by this message from their staff:

> These are very worrying times for all of us, and we find it difficult to put into words the feeling that we have for yourself, your family and the Brewery in general at this unfortunate time. We have all been thinking about you constantly but none of us wanted to add to your burden by individual expressions of support. We would therefore like to assure you that we are all aware of what you are going through on our behalf. We know that the past few weeks must have been the most traumatic and strength sapping of your entire life. Good luck and success in your negotiations to save the brewery and our jobs – no one could have done more. Whatever the outcome, may we wish you good health, peace and happiness when the present problems have been dealt with. You have earned it.

Then something remarkable happened. Batemans had been going to the annual Great British Beer Festival since it started, largely out of professional interest. The event, organised by CAMRA, was famous for two things: people drinking industrial quantities of beer from all over the country, and the awarding of Champion Beer of Britain. Jaclyn Bateman went to the 1986 festival, a particularly noisy event held at a Brighton hotel. She recalls:

> There was beer everywhere and it was very difficult to hear what was being said up on stage. Someone yelled at me that our XXXB had won, then everyone was cheering and that's how I found out! I was the only Bateman there as we had no inkling we would win – it was a great surprise. I have never drunk so much champagne in my life ...

XXXB had beaten off competition from 215 other beers from 120 breweries. It was a remarkable achievement and a brilliant opportunity to make Batemans better known, increase sales and make the company more investable. Luckily, Jaclyn managed to find a wholesaler in London – until then they had none outside Lincolnshire, making sales in southern England very challenging – and XXXB began to feature in pubs up and down the country. Further awards that year and after, at a regional and national level, also helped raise Batemans' profile. In fact, since that first win in 1986, barely a year has gone by in which they have not won something, an achievement that Batemans are enormously proud of.

Had the tide begun to turn? Maybe. It helped that George Bateman was the epitome of a country gentleman – smart, well spoken, affable and reasonable – and he was always happy to speak to journalists and started appearing in radio interviews too. Listening to one such broadcast was a London solicitor called Brian Bean, a senior partner in the firm of Paisner & Co., specialists in commercial property and corporate finance. Bean picked up his phone and called the Batemans; he wanted to help. 'There was a human element in our business that he liked,' Pat Bateman later recalled. 'We were so lucky. He took us up as a kind of crusade.' Stuart Bell, Batemans' long-serving company secretary, agrees his intervention was key. It helped that corporate legislation had been changed in 1985 to allow a company to buy its own shares from members and cancel them (a procedure called share buyback), which has become a very common way to assist shareholders in a private company who wish to leave the business. Bean managed to negotiate not only a sale price that the three parties could

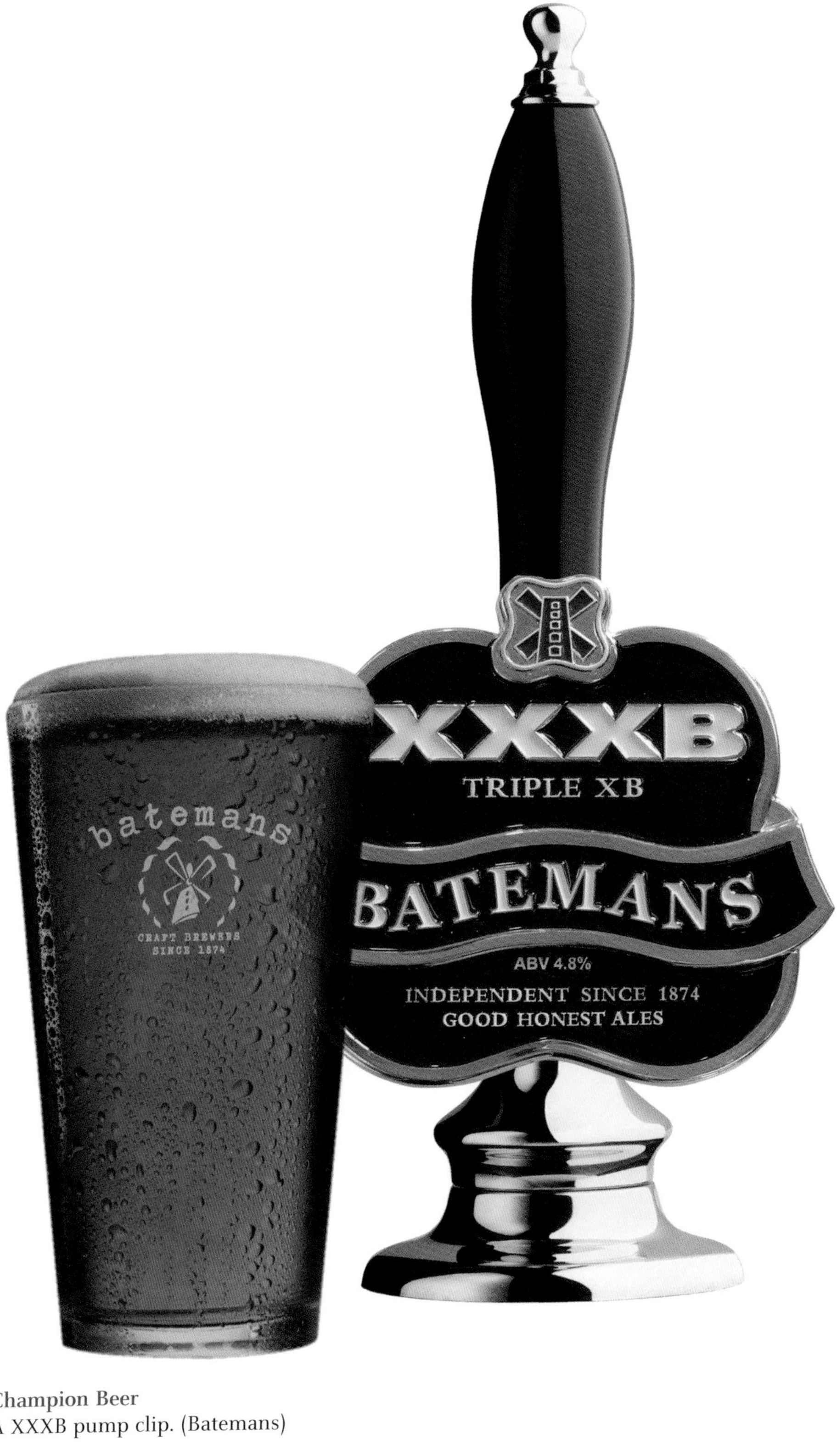

Champion Beer
A XXXB pump clip. (Batemans)

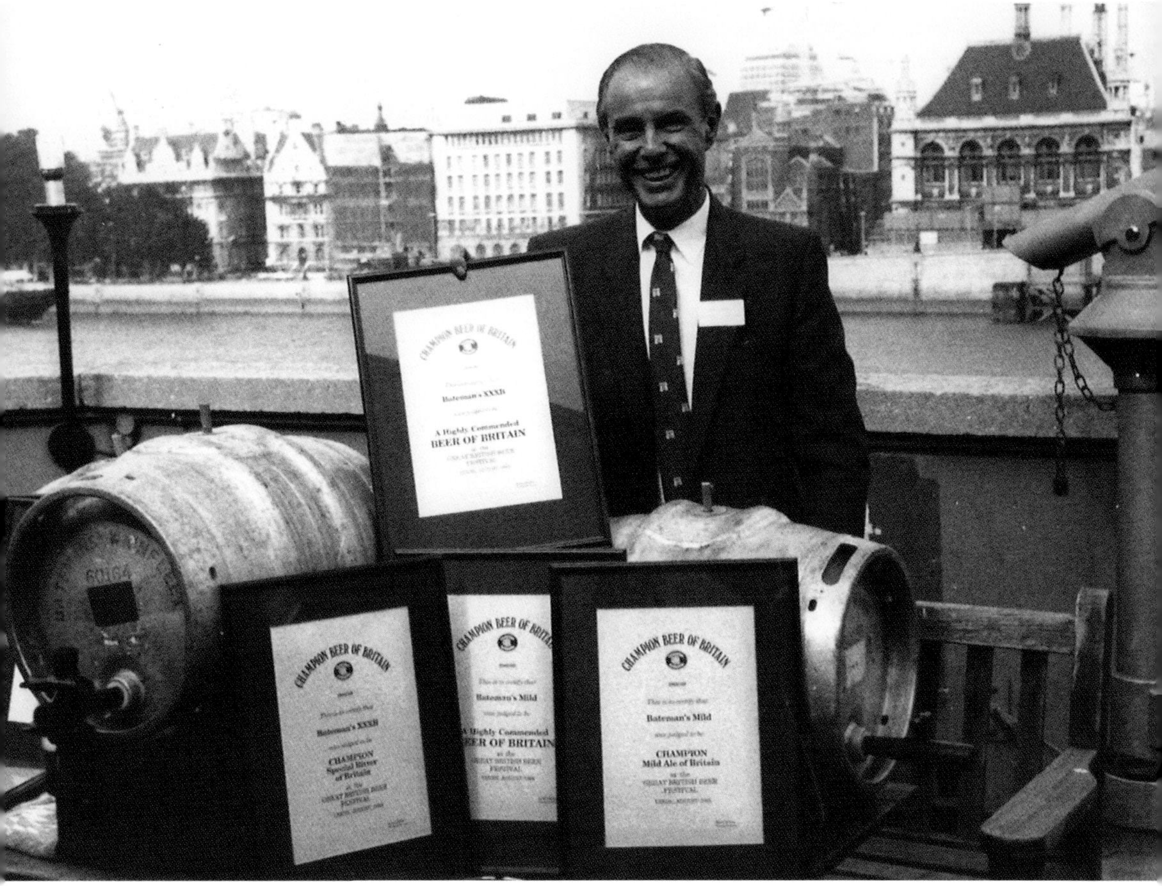

Hand on the Prize

A couple of years after Batemans won their independence, a very proud George Bateman, as always smartly turned out and sporting his brewery tie, is pictured with some of the awards Batemans received around that time. Clockwise from top: XXXB Highly Commended from 1988; Champion Mild of Britain, also from 1988; Mild Highly Commended; and XXXB Champion Special Bitter from 1987. The photograph was taken on the south bank of the Thames in London, near Blackfriars Bridge. (Batemans)

agree, but also one which banks were prepared to support, arranging for George to meet senior Midland Bank managers who agreed to loan the £3.6 million required.

On 30 January 1987, George Bateman and his family became the sole owners of the business. He quickly called a meeting to tell his staff, but the rumour mill had been revolving at speed and many of them knew what he was about to announce; in fact, the staff surprised George by presenting him with a crystal bowl engraved with the Batemans logo to mark the occasion. A vast banner was hung from the mill proclaiming 'V for Victory! Batemans will stay Batemans'. Helium balloons bearing the same slogan and in the brewery's red and yellow colours were handed out. 'People were cycling about with them on their handlebars,' Pat remembered. George said there was almost dancing on the streets of Wainfleet and ladies were stopping on the street to embrace him: 'It was all

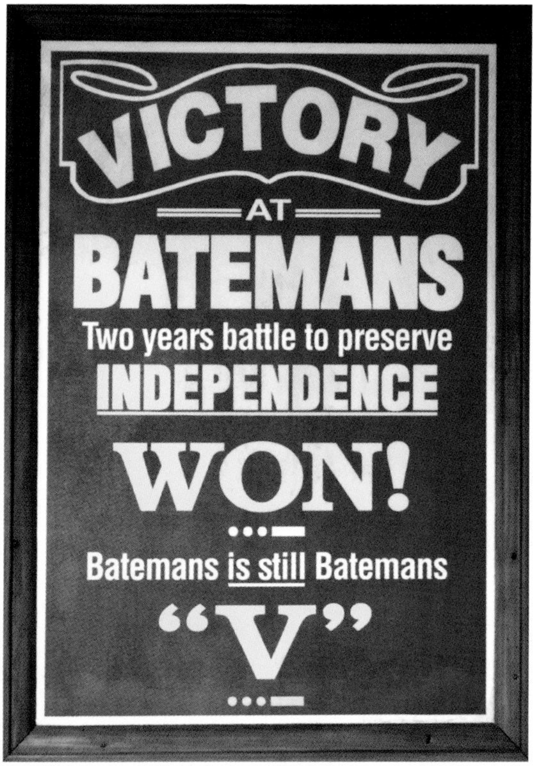

Left: **Independence**
This eye-catching poster in the house yellow and red colours was produced as part of the celebrations in 1987. The use of the Morse code 'dot dot dot dash' – meaning 'V' – is an interesting, though maybe unintentional, nod to Harry Bateman's clandestine radio operations in the Second World War. (Batemans)

Opposite: **Shared Celebration**
As part of the celebrations when XXXB was named Beer of the Year at the 1986 Great British Beer Festival, Batemans erected this enormously long banner on the main road into Wainfleet. The brewery staff – with the brewery and mill behind – cluster round the sign, with George Bateman (left) and head brewer Martin Cullimore (right) sharing a congratulatory handshake for the cameras. Jaclyn Bateman stands just behind her father. (Batemans)

rather splendid.' The following Saturday a surprise party for the family was arranged at one of their Skegness hotels.

Within a year, head brewer Martin Cullimore had perfected a celebration ale with the highly appropriate name of Victory, launched in April 1988 in both cask and bottles – Batemans' first cask beer to have a name rather than a series of letters. He remembers:

> I had a discussion with George about a beer to celebrate our independence, and suggested we should brew a stronger beer. George said 'what a great idea'. The process was I wanted a higher gravity pale ale brewed with traditional materials – not a stout or a dark beer. I used Maris Otter malt, 5% crystal malt and Goldings hops, and deliberately made it smooth. I made the mistake though of Victory being very drinkable. It was a 6% ale which drank like a 4% and that did have consequences for some drinkers.

The legendary beer writer Michael Jackson brought the TV cameras to Batemans the following year for his six-part documentary series *The Beer Hunter*, in which he travelled the world in search of great beer. The final episode was devoted to Britain, with Jackson calling in at Wainfleet to speak with George and family about the struggle for survival. Shown first on Channel 4 in December 1989, and then on the Discovery Channel in the US, the programme helped to raise Batemans' profile internationally.

To steer the company into what must have seen an exciting but still uncertain future, its corporate governance was updated, which saw Brian Bean and a local solicitor, Marcus

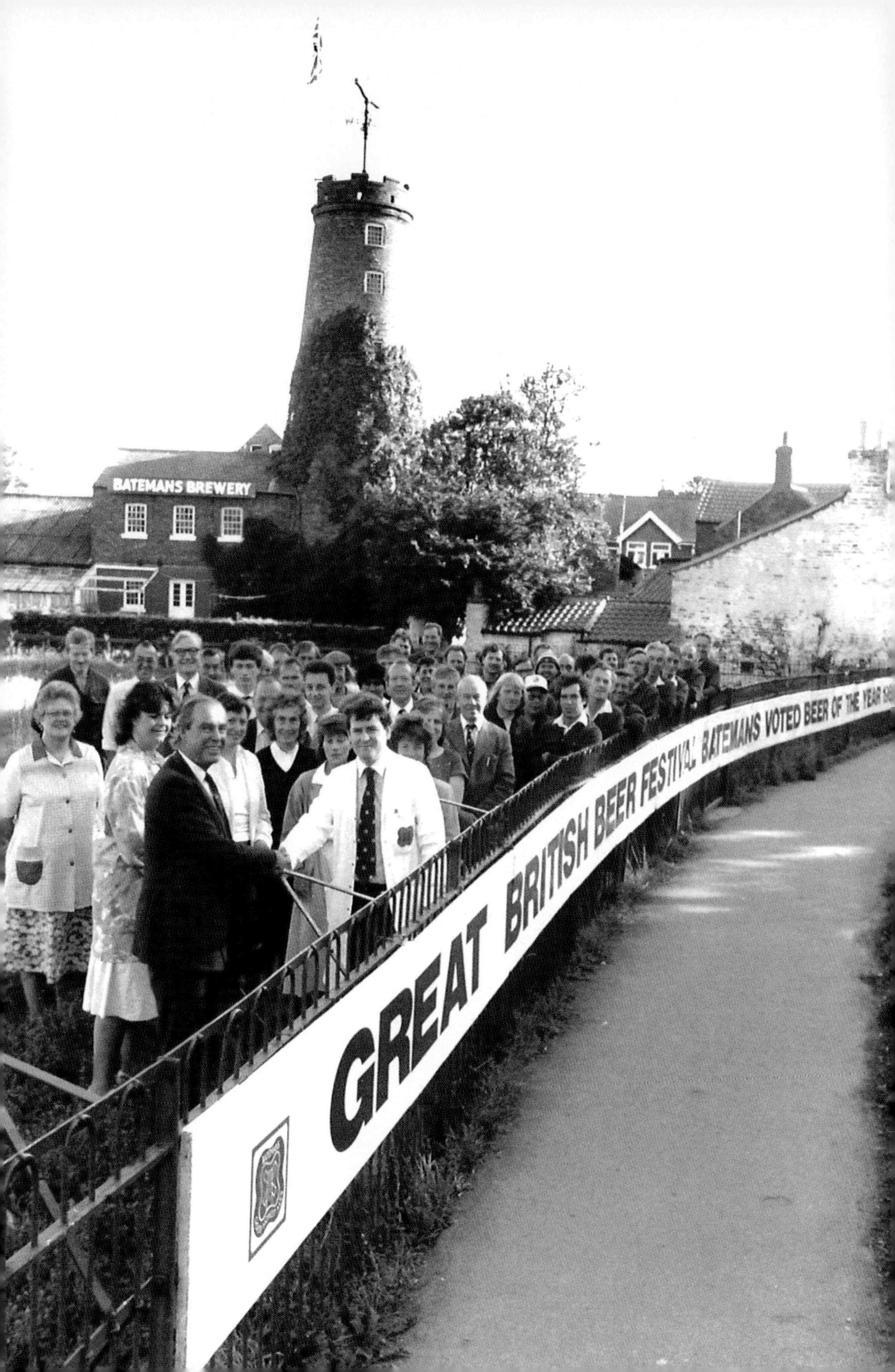

BATEMANS BREWERY
GREAT BRITISH BEER FESTIVAL
BATEMANS VOTED BEER OF THE YEAR

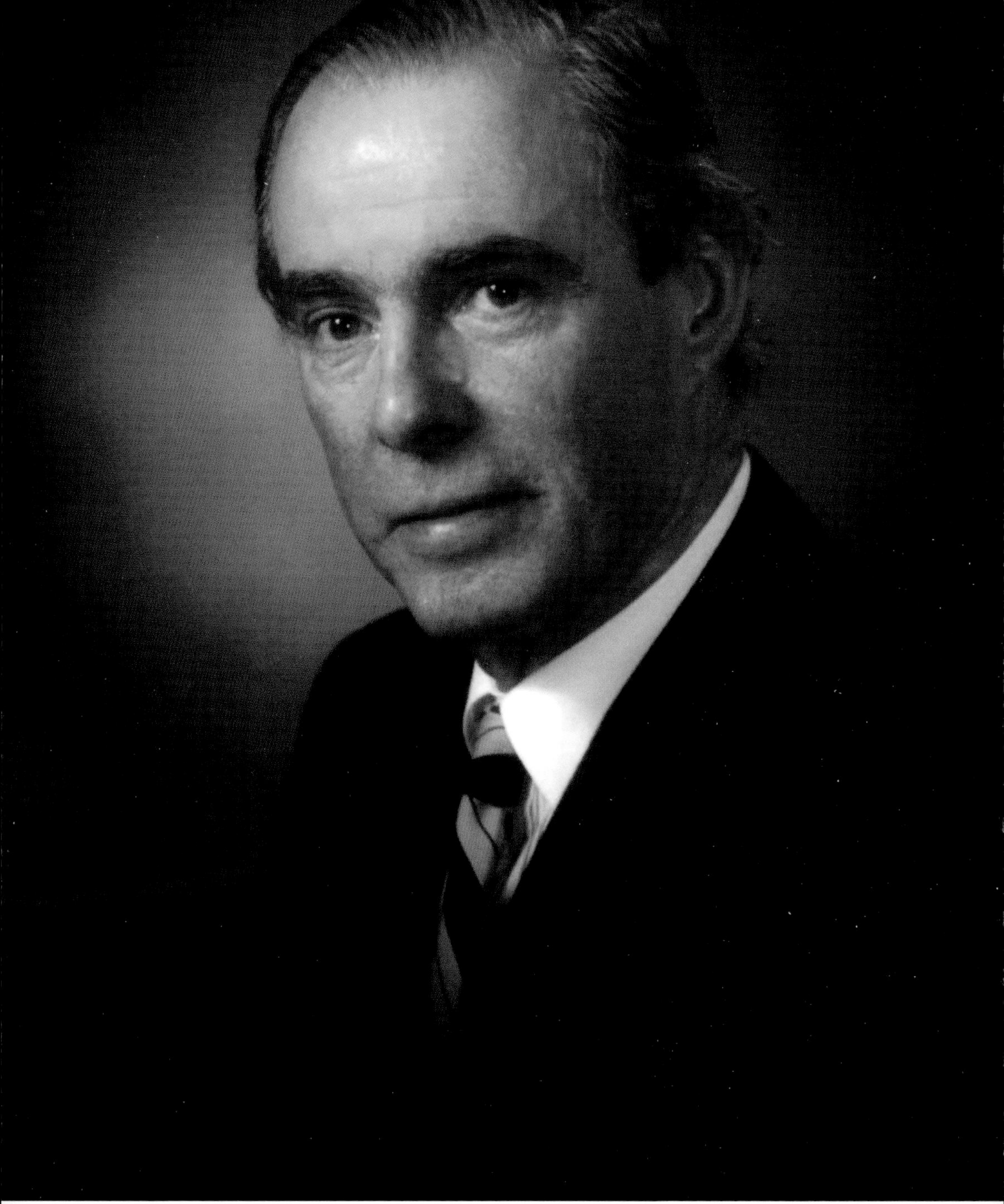

Head of the Firm
George Bateman pictured in a studio portrait from the 1990s. (Batemans)

Gunning, join the board, as well as Pat, Jaclyn and Stuart, who became assistant managing director. Changes were also made to the ownership structure of the business to avoid a repeat of the recent tumultuous events, placing the majority of the shares in trust for Stuart and Jaclyn. Stuart explains the reasoning:

> We never want to be in a position where we could endanger the future of the company and what it stands for – brewing good-quality cask ales, protecting local employment and pubs. Jaclyn and I agreed with our parents that in order to help ensure the continued independence of the brewery, a proportion of our shares should be held in trust, allowing us to control the future direction of the business, but not to trigger a sale.

But it was clear to all that independence had its price. The company coffers were practically empty, any available capital having been used to buy the shares held by other family members; repairs, maintenance and investment were virtually impossible. Even more worryingly, it became clear that many more of their pubs would have to go. George's original estimate that they might have to sell about ten seemed very optimistic. The estate was examined very critically, and within three years twenty-one mostly underperforming houses were sold, realising some £1.5 million; Stuart describes them as ones that would not cause too much local upset. It helped that the late 1980s property boom was in full swing, but even so some went for completely unexpected amounts, the most surprising being the Gate at Ulceby Cross, an unremarkable, though pleasant, roadside inn on the edge of the Wolds, selling around a barrel a week. It was put up for auction in February 1989 with a reserve of £35,000, but sold at £165,000 to widespread amazement. More sales took place until the end of 1994, by which time forty-five properties had gone – nearly half the entire estate – including all three Skegness hotels, the Vine, County and Parade. Some of the pubs had been Bateman houses for over seventy years, such as the Anchor at Friskney, and the Barley Mow a few miles to the south – a popular end point for the brewery tours they started running in the late 1980s. Some of them survived as free houses for some years afterwards; a smaller number still do.

One of the first tasks Stuart set himself when he became assistant MD was to renegotiate a number of existing trading agreements. The twenty-year agreement with Whitbread, signed in 1984, had not included any margin protection for annual price increases, so that Batemans were progressively worse off as each year passed; Stuart managed to get the agreement replaced with more favourable terms. This was followed by an agreement with Allied Breweries to transfer five pubs to Batemans, effectively for free on the understanding that Batemans would stock Allied brands in their pubs, including Ansells Mild, Carlsberg and Castlemaine XXXX lager (a defunct lower-strength version of an Australian beer licensed for the British market). Whilst Jaclyn worked with national wholesalers, persuading them to stock Batemans beers and sell them to supermarkets, Stuart developed a free trade operation, selling to pubs and clubs in Lincolnshire, Norfolk and the East Midlands. With the help of Basil Dixon who had been at Mansfield Brewery, Peter Vause, and Andrew Reed, who had been promoted from within, free trade became a major part of the business.

Batemans wanted to take advantage of their new fame and the cash generated by pub sales to expand outside their traditional Lincolnshire territory. In autumn 1987 they bought the George Inn at Ripley, Derbyshire, and a neglected former Shipstones pub in Nottingham, the Old Grey Nags Head, one of the city's first beerhouses. Part of the reasoning was that the

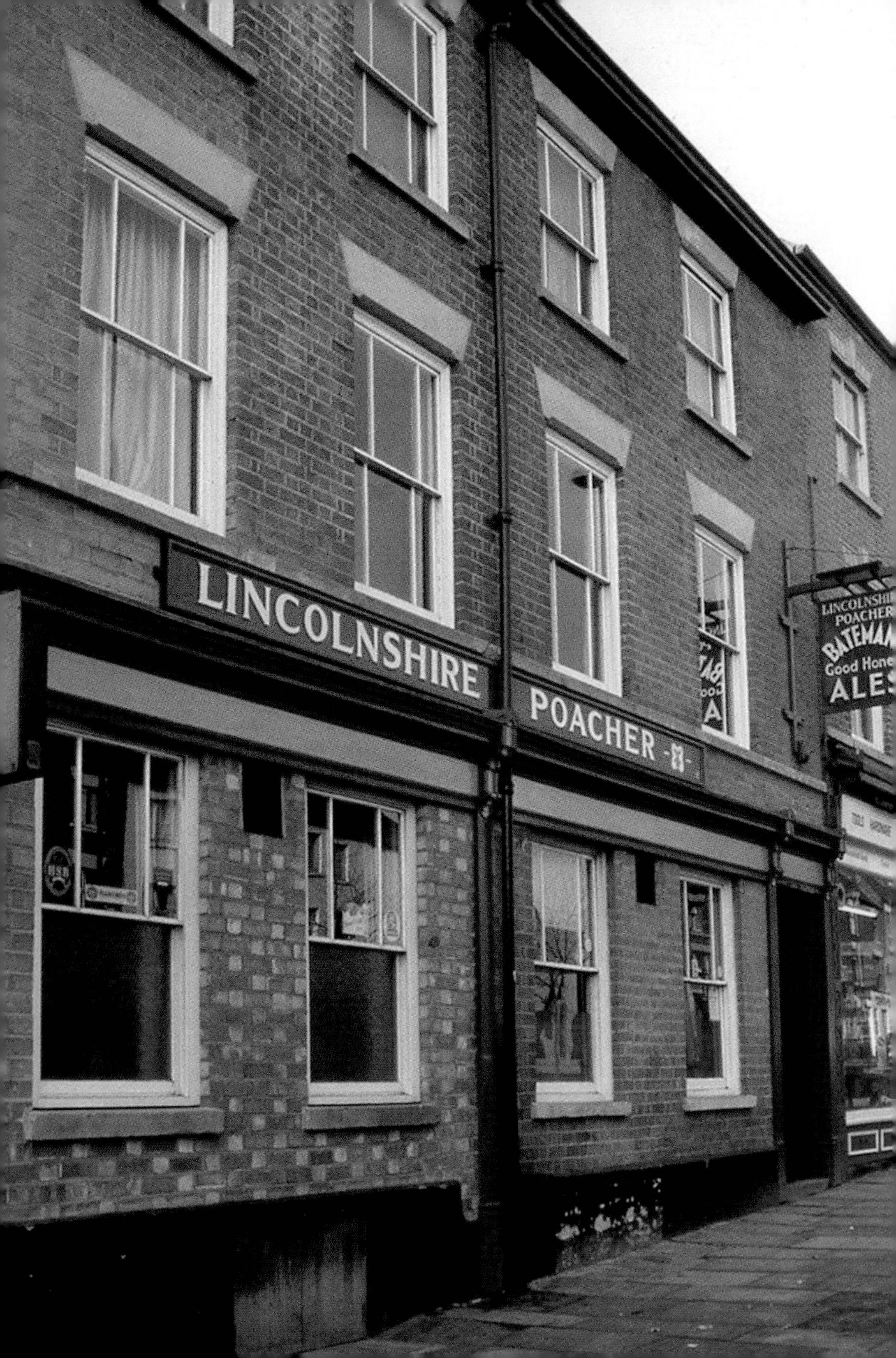
LINCOLNSHIRE
POACHER
POACHER
ALES

Bateman name was already known in the area from day trips to the Lincolnshire coast. The work needed at the Nottingham pub was substantial as it had been closed for two years, but eventually it reopened in June 1989, stripped back to basics as a real ale haven with a new name, the Lincolnshire Poacher. It was an almost instant success, trading at about twenty-five barrels a week (at a time when most Bateman pubs were averaging around four a week). The Poacher was an inventive joint venture with Tynemill, a pub company founded by former CAMRA chairman Chris Holmes. This was the first of a number of deals done between Stuart and Chris: when Batemans bought the Alexandra Hotel in Derby, a similar deal was struck for Tynemill to run it. Those early partnership agreements with Tynemill (now better known as Castle Rock Brewery) were the foundation for Batemans' flexible tenancy agreements, later adopted by many brewers who originally frowned on the initiative.

Opposite: **Total Transformation**
The Lincolnshire Poacher was Batemans' first house in Nottingham, reopening in 1989 after a thorough refurbishment. Formerly a Shipstones pub known as the Old Grey Nags Head, Batemans extended the premises into an adjoining shop and created a beer garden to the rear. It can be found on Mansfield Road and is now in the hands of Castle Rock Brewery, displaying a fascinating collection of beer and brewing memorabilia. One of the windows was etched with the original pub name until it was smashed recently. (John Law)

Below: **Branching Out**
The Bay Horse was a corner pub on Wincolmlee, a street in an industrial area of Hull, close to the river. Dating back to 1803, and rebuilt in 1878, Batemans bought it from Bass as part of their expansion outside Lincolnshire. It had a modern extension to the side, visible in this view, with a lofty grain silo looming in the background. The pub closed in 2017, a victim of declining custom, and has since been converted into flats. (Batemans)

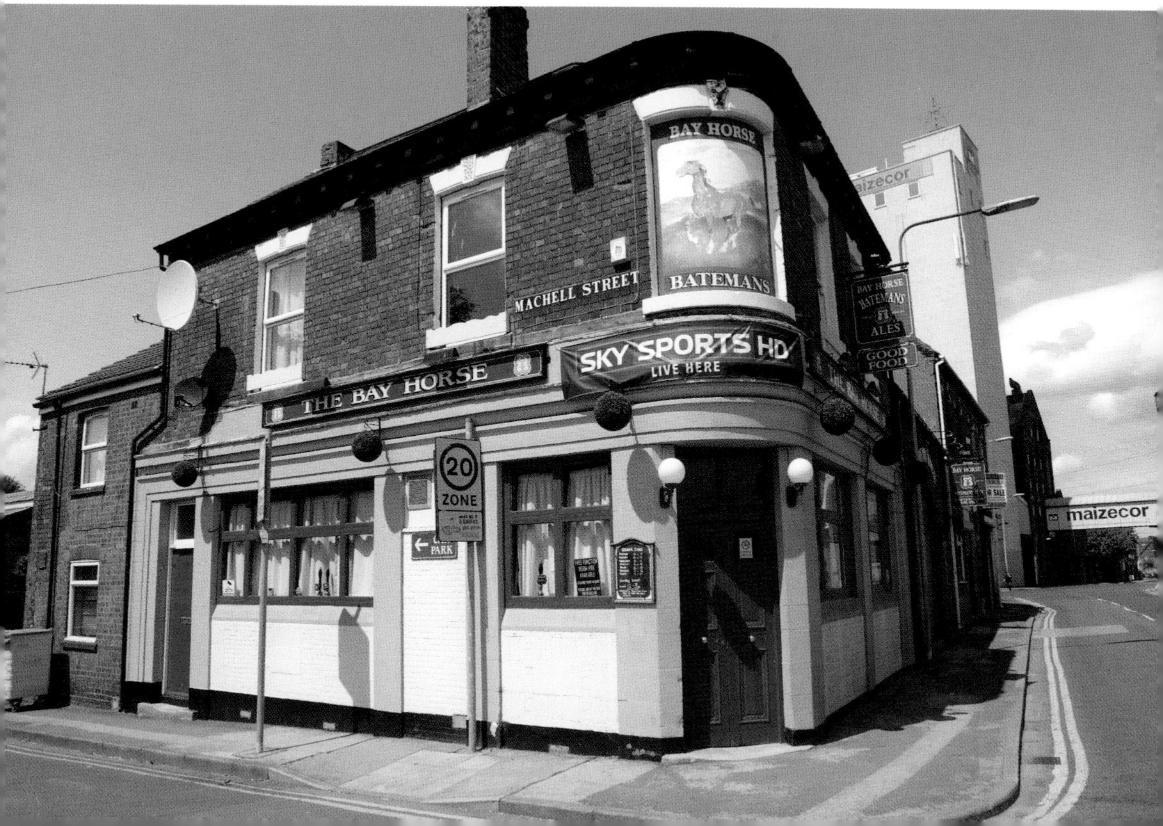

In an even bolder move, Batemans moved into London in 1987, snapping up the Nelson's Head in Bethnal Green, deep in the East End. This was at a time when several independent brewers felt they needed a London presence, but Batemans found it was simply too far away from Wainfleet and it was sold a couple of years later. Other purchases proved happier, including the Bay Horse in Hull and the Vine (later the Roaring Meg) in Newark, as well as a few more pubs in Lincolnshire towns, including the Woolpack in Wainfleet marketplace. Long coveted by Batemans, the former Home Brewery house was 'gifted' to them by Scottish & Newcastle in 1989, in a 'pubs for beer' agreement, in return for taking S&N brands such as McEwan's and Beck's lager. These inventive deals were hugely important at the time as Batemans would not have had the capital for a conventional purchase. A further branching out was the acquisition of Laws Drinks in 1993, a Grimsby-based soft drinks manufacturer and cash and carry business which Batemans had long traded with.

With this expansion, Batemans also looked at their range of beers and how to broaden their appeal to new customers. 'Concept' beers were becoming popular in the mid-1990s, challenging drinkers with new flavour combinations that they might not have thought of. Batemans introduced 'Mr George's Unique Flavours' in 1995, starting with 'Strawberry Fields Waynflete Bitter', which had fruit added in the hop back. The beer was very popular, but the fruit flavour was so strong that all other brews passing through the pipes tasted of strawberries, an issue only solved when someone suggested putting cider through the lines. A new pale ale flavoured with vanilla pods – which evolved into the present Yella Belly – proved more successful, followed by several more unlikely-sounding but popular recipes. Other beer ranges followed, three based upon the famous Skegness Jolly Fisherman: Jolly's Jaunt, Jollies Follies and Jollies Millennium. There was also the Mystic Brews range, the Calendar Girls, Good Honest T'Ales, and the Miss World beer ranges, all somewhat tongue in cheek. Of all those brews, one has stood the test of time very well: Rosey Nosey, Batemans' Christmas ale.

After the heady days of the 1980s, with large numbers of prospective tenants lining up at brewers' doors, the supply of suitable landlords fell away significantly. The problem came into sharp focus in 1989, when Batemans bought the White Swan at Stamford (sadly no longer a pub). For this attractive stone-built house, on a key road at the town centre fringe, there were only two applications. Stuart interviewed both of them:

> The first stated that he would not be using the catering kitchen as he didn't believe in food in pubs, he would not be having any sports teams and upon being asked what he thought of the local competition and what the trading opportunities were for the pub, he replied that he had never been round any of the other pubs. The other applicant came to interview with business plans, marketing plans, events calendars and sample menus. Obviously I gave the pub to him. However, he pulled out one week before the official opening and I had to take a chance on the first interviewee. Subsequently it turned out to be the best appointment I had ever made, whereas the second interviewee took another pub locally and was bankrupt within months. It just goes to show that interviewing tenants is no science!

On a much brighter side, Batemans made a key appointment in 1994, bringing in Haydn Biddle as chairman and chief executive. He brought the experience and precision of a career in senior roles at some of the country's largest businesses; he'd spent the last eight years at Scottish & Newcastle, then the country's fifth biggest brewer, winding up as managing director of their North East operations, but had also worked outside the beer world, as an advertising manager for the multinational consumer goods firm Procter & Gamble. He encouraged the family to appoint department heads so that responsibility and accountability was clearer, and recommended a more rigorous corporate branding exercise, which was led by Jaclyn; the Batemans logo started appearing on anything that could display it. But Biddle also understood the importance of a smaller family business like Batemans, and he remains chairman of Batemans to this day. He is also godfather to all three of Stuart's children. Another happy event of 1994 was Stuart's marriage to Rachael, who worked as a sales executive for the firm. Their wedding, of course, featured several toasts with Batemans' beer!

8

A Crowded Market, 2000–24

One of Stuart Bateman's first challenges when he joined the business had been to grow the free trade division – a sound decision, as it needed no capital investment. He explains that they concentrated on offering 'good beers, good people, great service and hard work'. It became a highly profitable and marketable asset, one that large corporates had been quietly eyeing up. In 2002, with Stuart newly installed as Managing Director, Carlsberg Tetley suddenly stepped in with an offer to buy the free trade for £3.6 million, as well as guaranteeing to take a minimum of 25,000 barrels of Batemans beer over six years. The only condition was that Batemans must not sell into the free trade during that time. Not only was this almost too good to be true, the offer was coincidentally the same amount as the 1987 bank loan, of which the huge interest payments continued to eat up all the trading profits. Now they could pay off all the loan, Batemans were finally debt free for the first time in many years. Stuart comments: 'Though it was a sad day when we sold our free trade, that six-year period was soon over and when we were able to sell to the trade again, we built the division back 30% bigger than before.'

Another line which Stuart was keen to develop was supermarket sales. It was a market that had grown slowly since the early 1960s – before then, supermarkets weren't allowed to sell alcohol – but by the 1980s the picture was very different. Ruddles, not far away in Rutland, had gone so far as to sell all but one of their pubs back in 1978 to concentrate on supermarket sales, starkly demonstrating what a lucrative market it could be. Although Batemans were selling some beer to supermarkets, the opportunities created by selling off their free trade allowed them to broaden their national sales division, which grew from 2,500 barrels in 2003 to 13,000 in the space of five years, and becoming the sole supplier of premium bottled beer to Aldi. The range and presentation of bottled beers was updated, with Victory, XXXB, Dark Lord (a bottled version of another new beer, Salem Porter) and Combined Harvest (a multigrain beer), allied to a number of seasonal lines, widely available in shops during the 2000s. This was a time when major supermarkets were keen to run 'beer challenge' events, in which several bottled ales – often specially produced new beers – would compete against each other for a limited period on the store's shelves. The winners would be rewarded with a substantial order. Batemans' biggest success came in 2002, when they were up against much larger brewers in the Tesco Beer Challenge. They were very proud to win with Harvest Zest, a grapefruit beer since retired. Mocha, a beer flavoured with coffee, won a similar competition at Sainsbury's and is still part of the range.

Batemans returned to contract bottling as well, something not seen since the early 1970s when they ceased to bottle Guinness. The most outlandish contract was to brew and bottle 'Mad Eddie's Chilli Beer', a lager with a chilli floating in it. The brand owner couldn't find any other packager who could get the chillis into the bottle. Adhering to George Bateman's dictum of 'always say yes and then work out how you're going to do it', rather than invest in a hi-tech machine, Stuart got one of the bottling store lads to pop a chilli in each bottle as it went past on the conveyor. However, he decided not to wear the protective gloves he had been given and by lunchtime he was in hospital, as the chillis had been absorbed into his skin, resulting in massive swelling.

Equally as successful as the off trade was the national on-trade, which over the same period grew from 2,500 barrels a year to 15,000, through a chain of national wholesalers including The Beer Seller, Tavern and Matthew Clarke. Independent wholesalers were important too, such as Small Beer of Lincoln and Dayla of Aylesbury. One of the early successes for Batemans' National Sales team was picking up a contract to be the sole supplier of Dark Mild to Marstons, which was worth over 2,000 barrels a year.

One project that didn't fit George's philosophy was a request to bottle an alcoholic lemonade from Australia called Two Dogs. Having said 'yes' Stuart then worked out that what they wanted bottling in a day would normally take three months through the Batemans plant. Eventually the levels of production that were being attempted became too great for the bottling plant at Wainfleet, a rambling place of clattering bottles and unexpected pipework ready to trap unwary heads. It was a marriage of second-hand machinery from a soft drinks company and equipment rescued from a scrapyard, efficient for basic bottling, but as bottle presentation became more important, the plant could not offer the flexibility needed, nor could it cope with palletisation. The investment needed was just too much, and bottling was therefore outsourced to Marstons of Burton-on-Trent.

Existing supply deals were reviewed. Stuart remembers meeting with one well-known national brewing conglomerate:

> We had signed up to an agreement with them some years ago to take over five of their pubs and pay for them in instalments over seven years, in return for which we had to take 4,000 barrels a year of their lager for twenty years. It had all been a bit Big Brother really. Although this was a twenty-year agreement, the economics of it only really worked for us for a period of seven years, due to the fact that the interest we saved was offset by the price increases which were put through by the supplier on the 4,000 barrels per year, meaning Batemans would be well out of pocket for the remaining thirteen years. After lengthy discussions, they agreed to rebalance the agreement, which saved us thousands and thousands of pounds. This may have been aided by the fact that it transpired that the national brewer had lost their copy of the original agreement and weren't quite sure what it had said!

Sales were quite buoyant in the early 2000s, helping Batemans to win a number of awards. Riding on the crest of that wave, they devised a 'Lincolnshire Comes to London' event, taking over a pub in the capital for the day and inviting customers from all over the

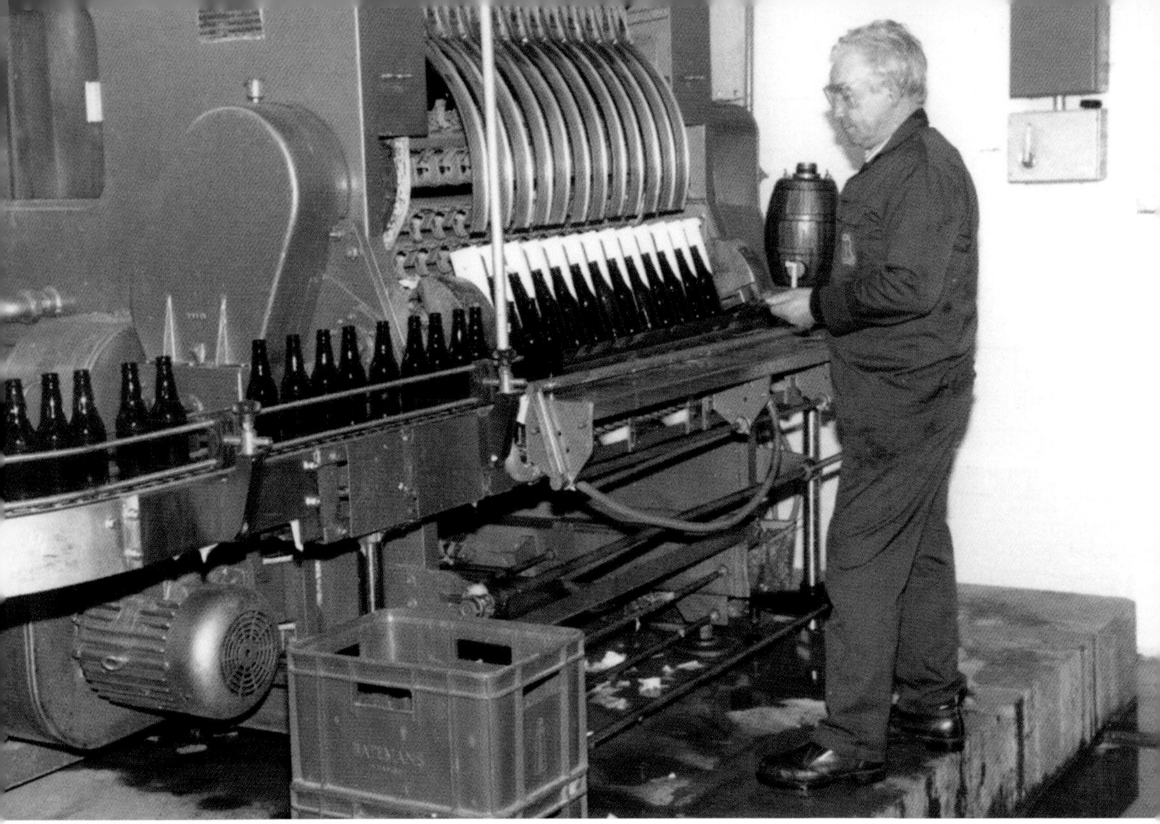

Lot of Bottle
Ray Symonds is seen at work at the bottling plant, which was updated in the 1960s by brewery engineer Peter Sharpe; eventually bottling was contracted out to Marstons of Burton-on-Trent. Ray and other family members racked up nearly 220 years' service with Batemans. In the foreground are a couple of Batemans' bright-red plastic crates. Unexpectedly, an episode of the BBC soap opera *EastEnders* featured Phil Mitchell sitting on one in the Queen Vic. (Batemans)

country to celebrate the latest victory. Another promotional event from that time – which still takes place annually at Seacroft Golf Club in Skegness – demonstrates Batemans' ethos of not taking themselves too seriously and believing that working with pubs, beer and people should be fun. Stuart explains:

> Our annual Golf Day became one of the most popular events on the industry calendar. In line with our general philosophy, we did not take ourselves too seriously – we called it 'Bit of a Bash by the Beach'. It's had various themes: Elvis Presley, road workers, leprechauns, football, Donald Trump, Germany, and even a circus, at which there were clowns and jugglers wandering around the course. That one ended with me being shot out of a cannon for charity!

The success of new products had at last allowed serious capital investment, in the shape of a new brewhouse, allowing production to reach 35,000 barrels a year, but also the flexibility to brew small-batch beers. Project managed by head brewer Martin Cullimore, the state-of-the-art facility was opened with great ceremony in March 2002 by the well-known beer writer Roger Protz, whose articles over the years had done much to help raise Batemans' profile. Rather than adopt the unadventurous name 'the new brewhouse', a much more original – and appropriate – name was chosen, 'The Theatre of Beers', which is a play on the Manchester United's 'Theatre of Dreams' moniker for their home ground. Stuart came up with the idea of giving people the chance to sponsor a brick engraved with their name on the inside wall of the brewhouse, an opportunity taken up eagerly by 180 beer enthusiasts who were invited to the opening ceremony. George Bateman spoke movingly of how proud his father Harry would have been to see the 'Theatre of Beers' helping to secure local jobs for the future.

A new visitor centre was opened in 1993 in the former bottling room – a surprisingly large space with all the plant removed. The opening ceremony was performed by veterans from 617 Squadron, Royal Air Force – the famous 'Dambusters' who'd flown from nearby airfields, to whom Batemans ales would have been a welcome diversion in wartime. It was also the 50th anniversary of the Dams raid. The lack of proper entertaining facilities was just part of an event Jaclyn Bateman won't forget in a hurry:

> At that stage we had no kitchen, so we entertained them in the bottling store, which was then still in use. We had to cook them a meal, and for the first course I made

Grand Opening
Beer writer Roger Protz opened Batemans' new brewhouse, the Theatre of Beers, in March 2002. He is seen here shaking George Bateman's hand during the ceremonies. Martin Cullimore, in his head brewer's white lab jacket, is on hand to the far left. The inspiring name is a take on the 'Theatre of Dreams' name given to Manchester United's home ground – Batemans were keen to avoid just calling it 'the new brewhouse'. (Batemans)

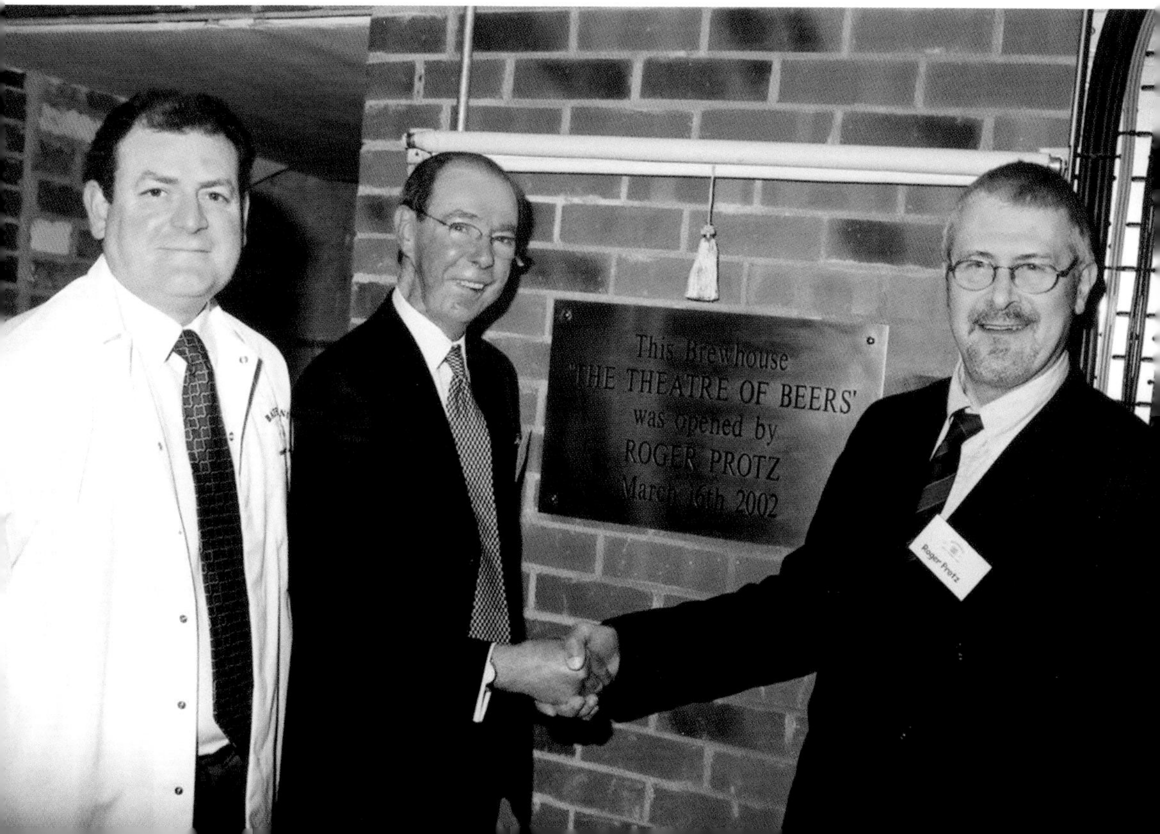

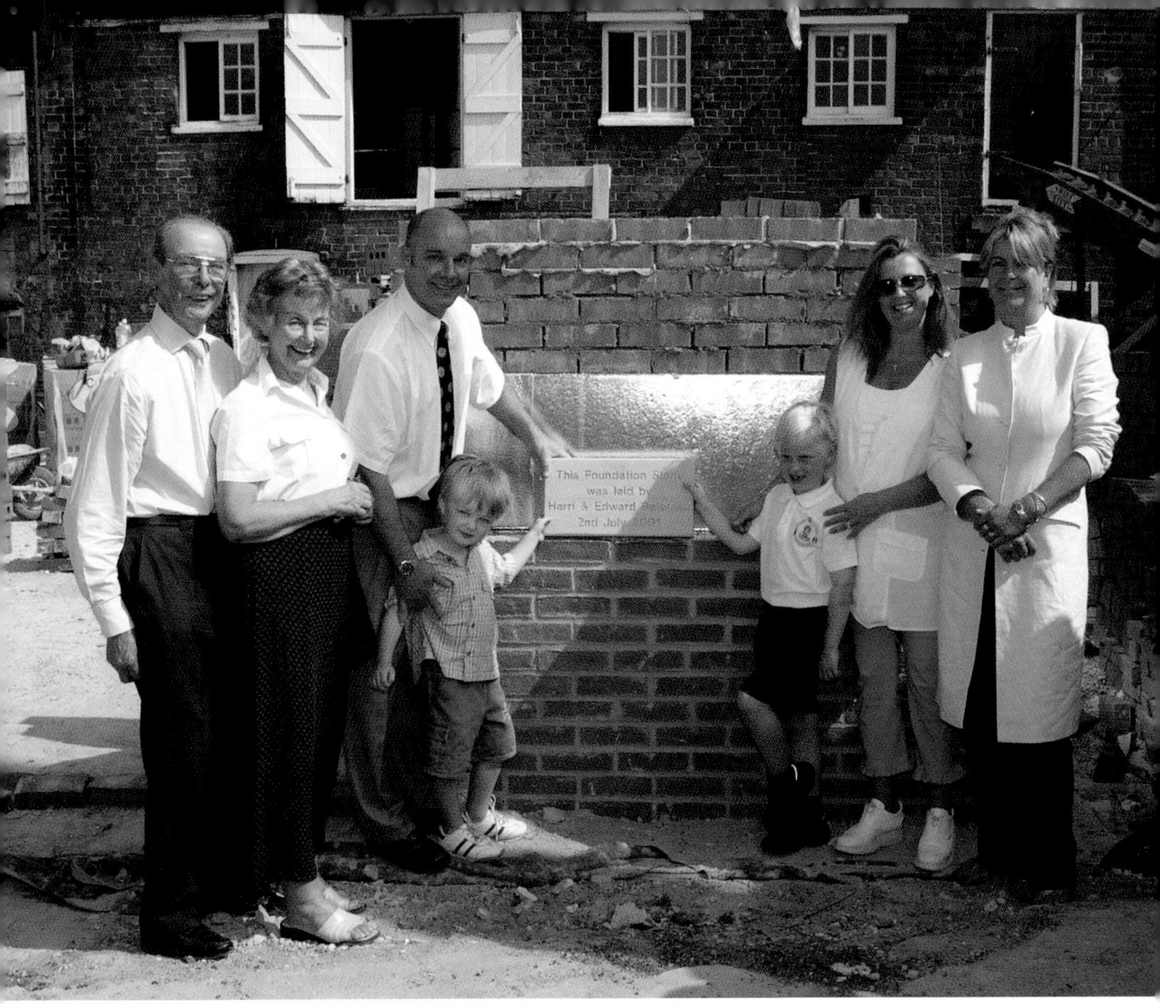

Family Gathering
Three generations proudly mark the laying of the foundation stone for Batemans' new modern beerhouse, 'The Theatre of Beers', in July 2001. From left, the Batemans are: George, Pat, Stuart, Edward, Harry, Rachael and Jaclyn. (Batemans)

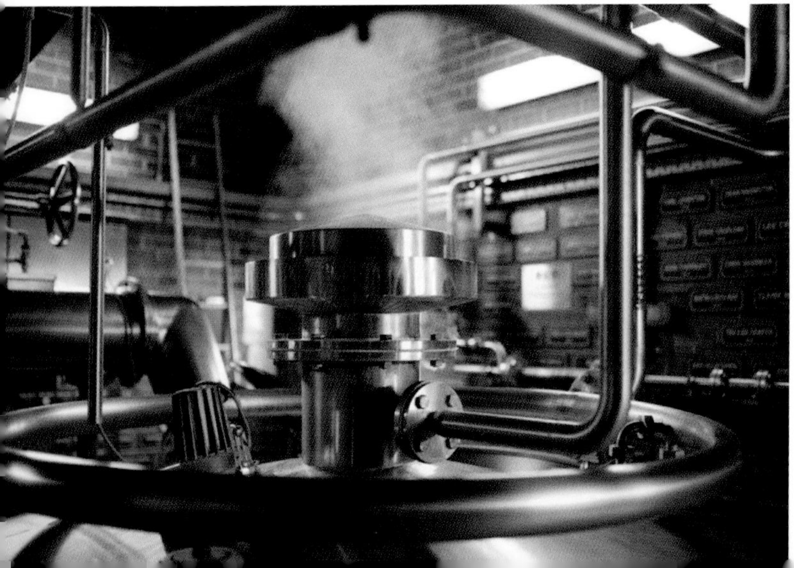

Full Steam Ahead
The stainless-steel mash tun in The Theatre of Beers, opened in 2002, was installed by Shobwood Engineering of Burton-on-Trent. Here it is at work. On the wall behind are the inscribed bricks of Batemans' Beer Club, each bearing a member's name. (Batemans)

> vichyssoise with fresh Lincolnshire leeks, using six huge tubs that our sugar was delivered in. I added some Oxo cubes, but the yeast in them reacted with the sugars in the vegetables and everything started fermenting. I had to get rid of the lot which wasn't easy with six tubs of the stuff. I had to pour it down the drain and start again ...

Richly decorated with posters and memorabilia, the centre houses a vast collection of around 5,000 bottled beers, probably the country's biggest, which George Bateman had started back in 1954. The mill was incorporated into the centre, with a mini pub, named Mr George's Bar, on the ground floor, an artifacts room above, and on the next floor up a recreation of Harry Bateman's office as it would have been back in his heyday. All the rooms are circular, following the shape of the mill tower, so that you can hear echoes all around the bar.

Although George and Pat were enthusiastic participants at these grand events, sadly George was not in the best of health. Soon after leading the negotiations to secure Batemans' independence, he'd contracted cancer of the mouth, requiring an operation which affected his speech. He could have chosen to retire, but he returned to work. Further operations followed, and then the cancer worsened. He died aged seventy-nine on

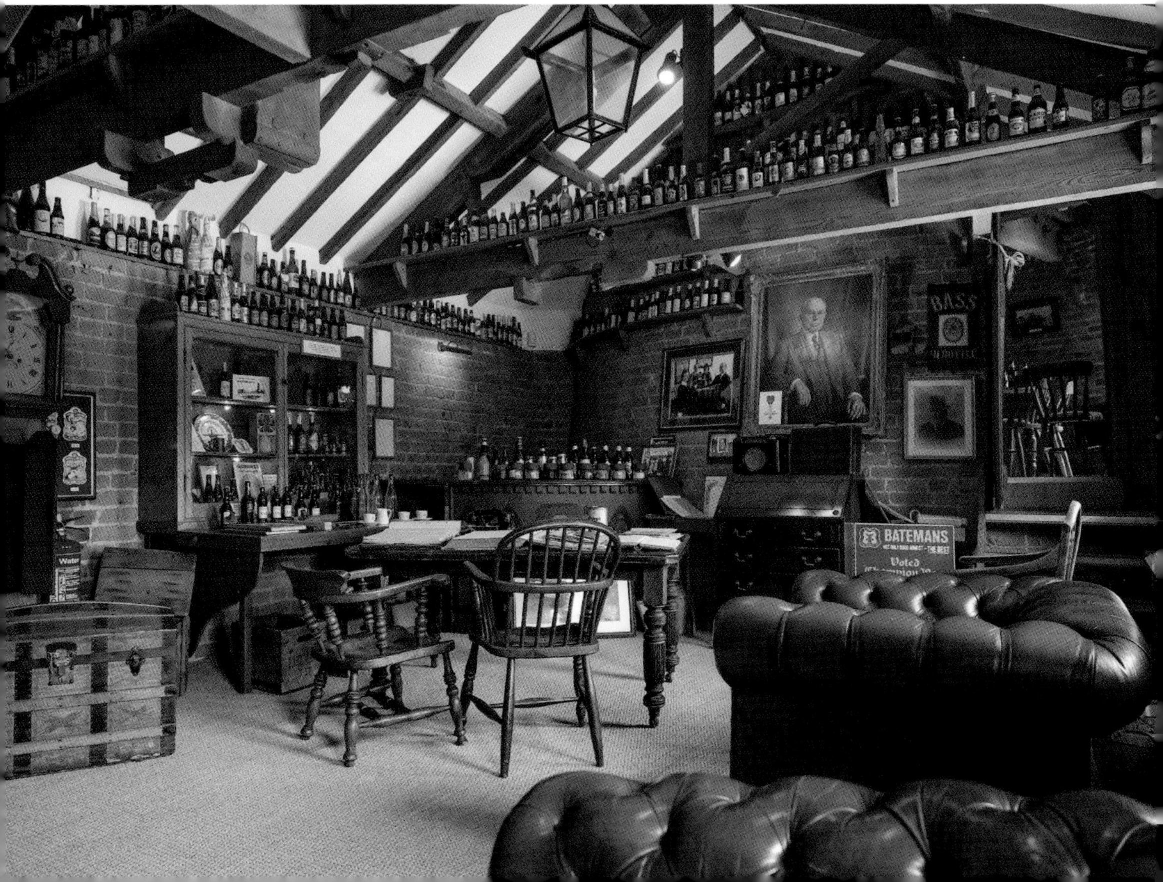

Museum Piece

The artefacts room at the brewery, with the curved wall of the mill in the background. The collection of bottled beers was started by George Bateman and continues throughout the visitor centre. It's believed to be one of the biggest in the country. Harry Bateman's portrait hangs on the mill wall. (Batemans)

25 June 2007, though Pat had predeceased him a couple of years earlier. Like his father and grandfather, George had seen his role as more than just a brewery director; he was one of the founders and the first chairman of Linkage, a local charity for people with learning difficulties, as well as a host of other voluntary posts. Roger Protz spoke at his funeral: 'He was funny, convivial, warm and generous. He was a legend, an iconic figure. If he had critics or detractors, I never met them. People simply loved George. When he fought to save his brewery in the 1980s, everyone – beer lovers, publicans, even rival brewers – rallied to his side.'

On a much happier note, Batemans continued to win top awards, and not just for the quality of their beers. The standards they set in their pubs also started to be recognised, culminating in 2009 when they won the prestigious Publican award for Pub Company of the Year – the Oscars of the pub trade – followed in 2010 by Regional Brewer of the Year. As some of the competing pub companies and brewers have much bigger estates and greater resources, this success was a clear demonstration of Batemans' ability to punch above their weight. To help incentivise tenants and reward hard work, Batemans decided to start their own recognition programme, the Rewarding Success Awards. There are a variety of categories, culminating in Publican of the Year. The first winner, back in 2010, was the Butcher and Beast at Heighington, just south of Lincoln, a charming stone-built pub famous for its impressive floral displays.

Unique Name
There are plenty of Butchers Arms up and down the country, but the Butcher and Beast at Heighington, near Lincoln, is the only one that acknowledges a key part of the butcher's trade. Bought in 1967, it's constructed in a mix of local stone and yellow brick. (Batemans)

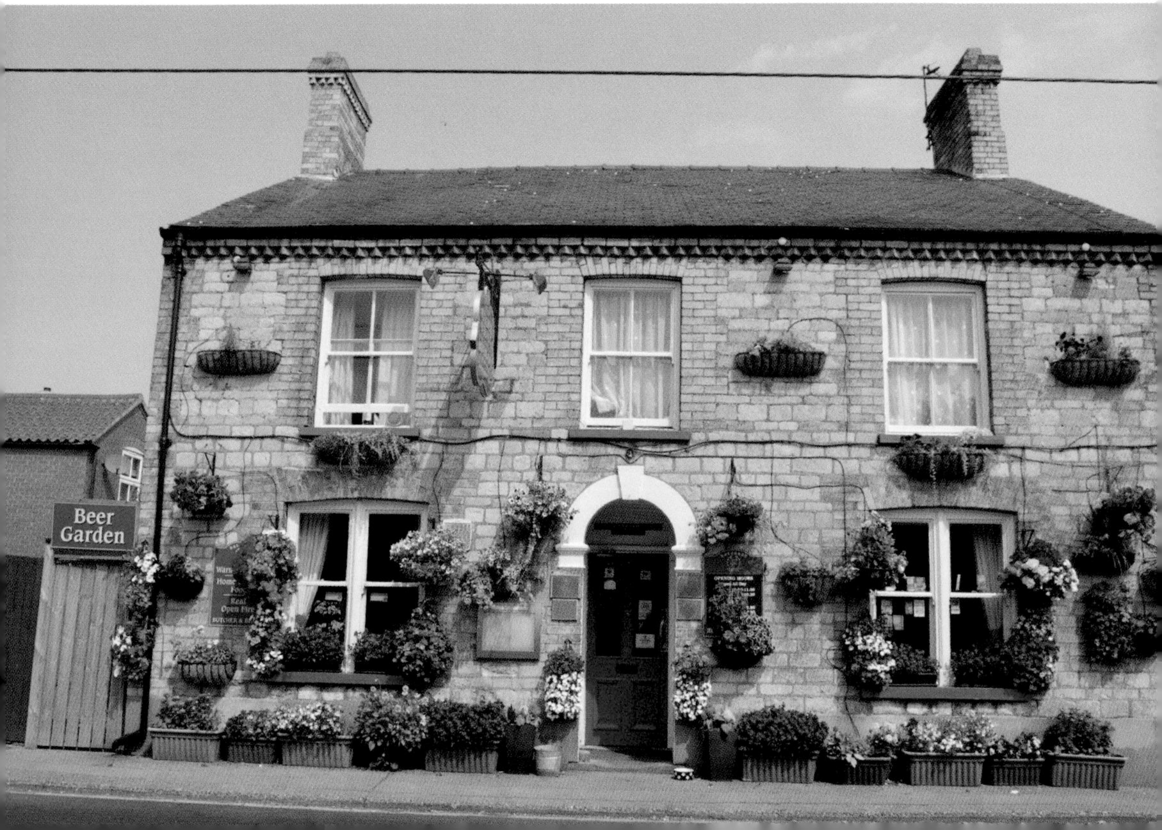

These days, Batemans are competing in a very busy market: the number of UK brewers continues to surge past 2,500, compared to a low point of just 147 in 1976. The popularity of craft beer is behind this extraordinary growth (though overall production is decreasing). The range of beers available from these new independents is incredibly wide; if you want your beer to be very 'hop forward' or have notes of grapefruit or mango, or all of those things, accompanied by some funky branding, maybe served in a bar with an industrial warehouse vibe, then you could be spoilt for choice. Whilst great news for drinkers, it placed established brewers under some pressure to respond. For Batemans' part, they introduced a wide range of beers, with increasingly clever marketing, both in bottled and draught form, and updated the brewery's windmill logo, which was starting to look a little dated. Presentation of the bottled beer range was given a thorough upgrade, replacing the old 'wavy label' with a sharp, modern look – as Stuart says, attractive bottle labels are key to creating an initial visual pull for drinkers. And a brace of new straplines appeared: 'Beer Brewed by Brewers, not Accountants' and 'Craft Brewers Since 1874'. There have been many definitions of 'craft', but Batemans consider the most accurate one is 'skills passed

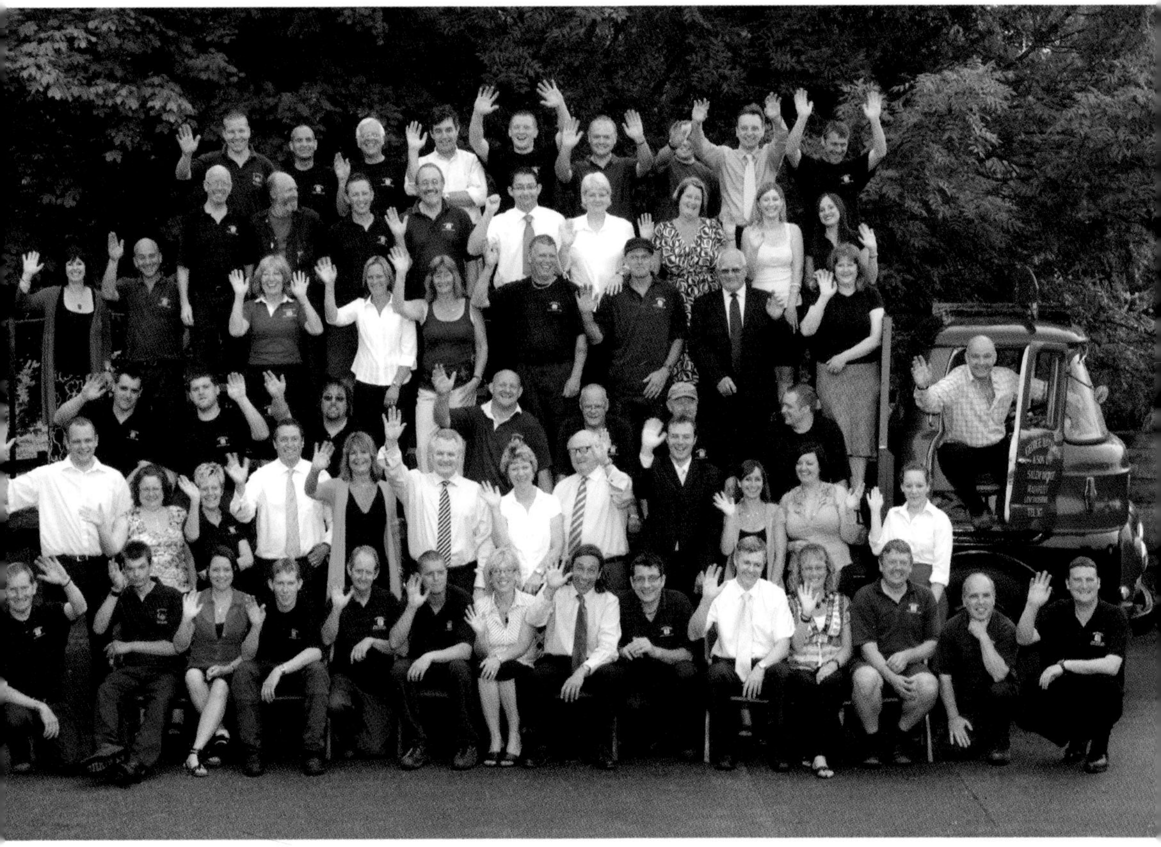

Happy Family
Most of the 2009 payroll gather around – and on – the company's heritage brewery dray. There are too many to name individually, but Stuart Bateman is in the lorry's cab and Jaclyn is fifth left in the second row. (Batemans)

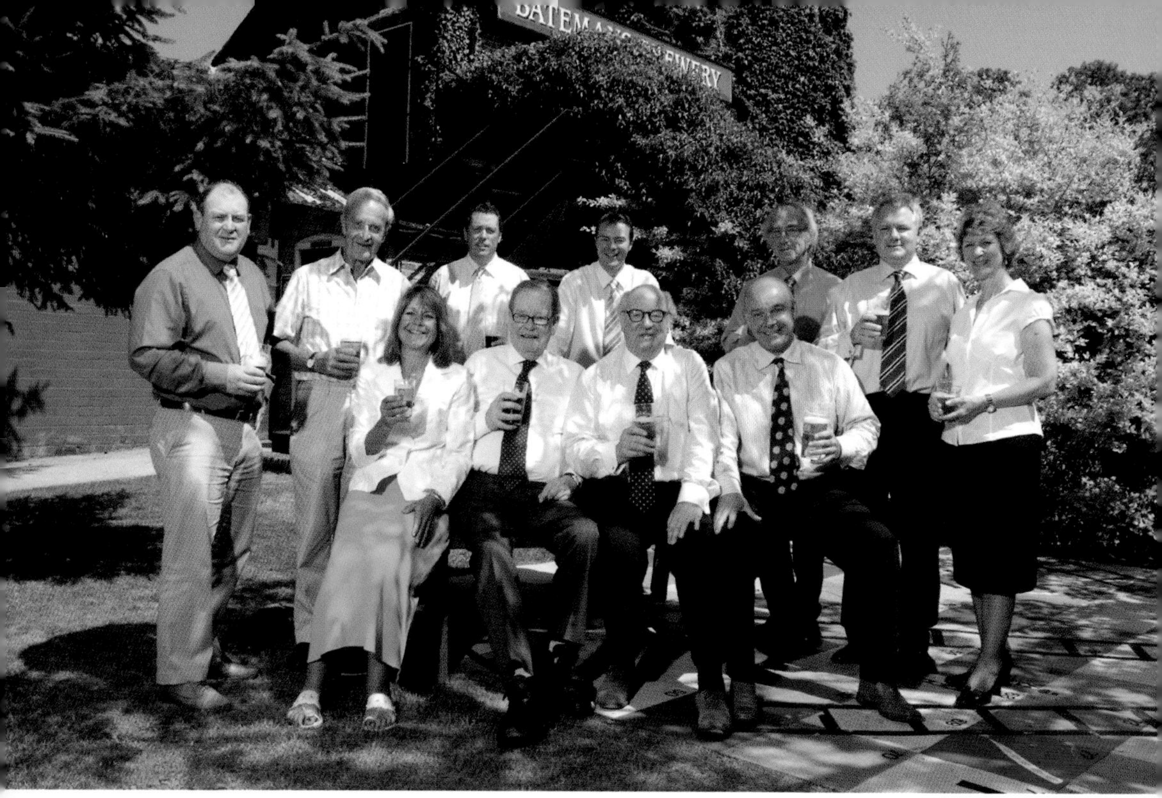

Steering Group

A group photograph from 2009 of the company directors, taken on a gorgeous summer's day in the mill garden. Top row, from the left: Martin Cullimore (Head Brewer), Peter Hammond (Non-executive Director), Andrew Reed (Sales Director), Rob Jones (Pubs Director), John Else (another non-exec, formerly in charge of Stuart's training at Mansfield Brewery), John Woodward (Finance Director) and Sandra Raymond (Secretary). Bottom row: Jaclyn Bateman, Brian Bean (Joint Chairman), Haydn Biddle (Joint Chairman), Stuart Bateman (Managing Director). (Batemans)

down from generation to generation' and with five generations of Batemans having been involved in the brewery, each learning from the previous generation, they certainly qualify.

There was a further factor at work, this time impacting upon finance and production. HM Treasury had introduced Small Brewers' Relief (SBR) in 2002 aimed at supporting small and emerging firms by allowing them to pay lower excise duty on beer production, and thereby compete on a more level footing against global brewers. However, when annual production exceeded 5,000 hectolitres, the amount of duty relief steadily declined, although some reduction was still available up to 60,000 hectolitres. Batemans were producing close to that ceiling when they were brewing about 50,000 hectolitres a year in the early 2000s and found themselves in a strange position – like other similar-sized brewers – that the SBR ceiling was driving them to cut production to benefit from lower duty rates. This was mainly why Batemans pulled back from several national contracts in 2014, making their beers less easy to obtain outside their local markets in Lincolnshire, the East Midlands and East Anglia. The way that the duty relief worked was adjusted in August 2023, so that it's

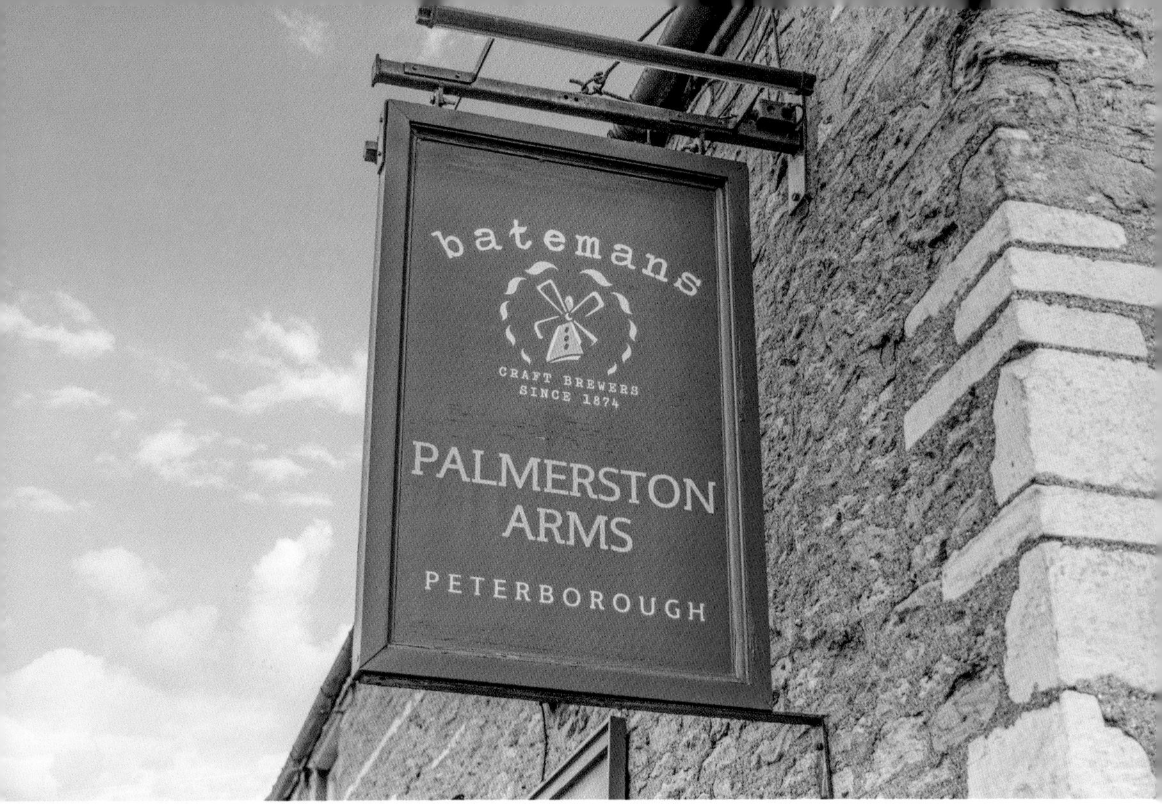

Lower Case
Mid-2010s branding at the Palmerston Arms in Woodston, Peterborough, with a lower-case font, stylised windmill logo and the 'Craft Brewers Since 1874' strapline. (Batemans)

viable for Batemans to increase the volume of beer brewed, partly by the purchase of more pubs and partly by putting more resource into their free trade operations.

As of October 2023, the estate stands at forty-nine pubs, most of them tenanted under Batemans' Profit Partnership scheme, which provides for a fairer partnership between brewer and tenant. It's a far cry from the 'wet rent' system that still persists, which results in tenants paying heavily inflated prices as a form of additional rent – hardly a relationship based on trust, as Stuart points out. Instead, Batemans scrapped all rent reviews, introduced a fair and transparent pricing structure, and made the cost of taking on a pub far more realistic and affordable. These reforms cut through decades of poor practice in the pub industry. More recently, Batemans have introduced the Reduced Risk Business Purchase Agreement, which allows the tenant to buy the goodwill of the pub business. There isn't one standard type of Batemans pub either. Yes, they still have traditional country and town pubs, but others cater to a very different market, such as the former Britannia in Boston, now a party pub. Open three days a week, closing in the small hours on Fridays and Saturdays, it's been rebranded as the Brit Bar ('your one stop, non stop party zone').

The Covid-19 outbreak that started in early 2020 affected all areas of life. For the pub trade, in common with much of the wider retail sector, there were periods when pubs were

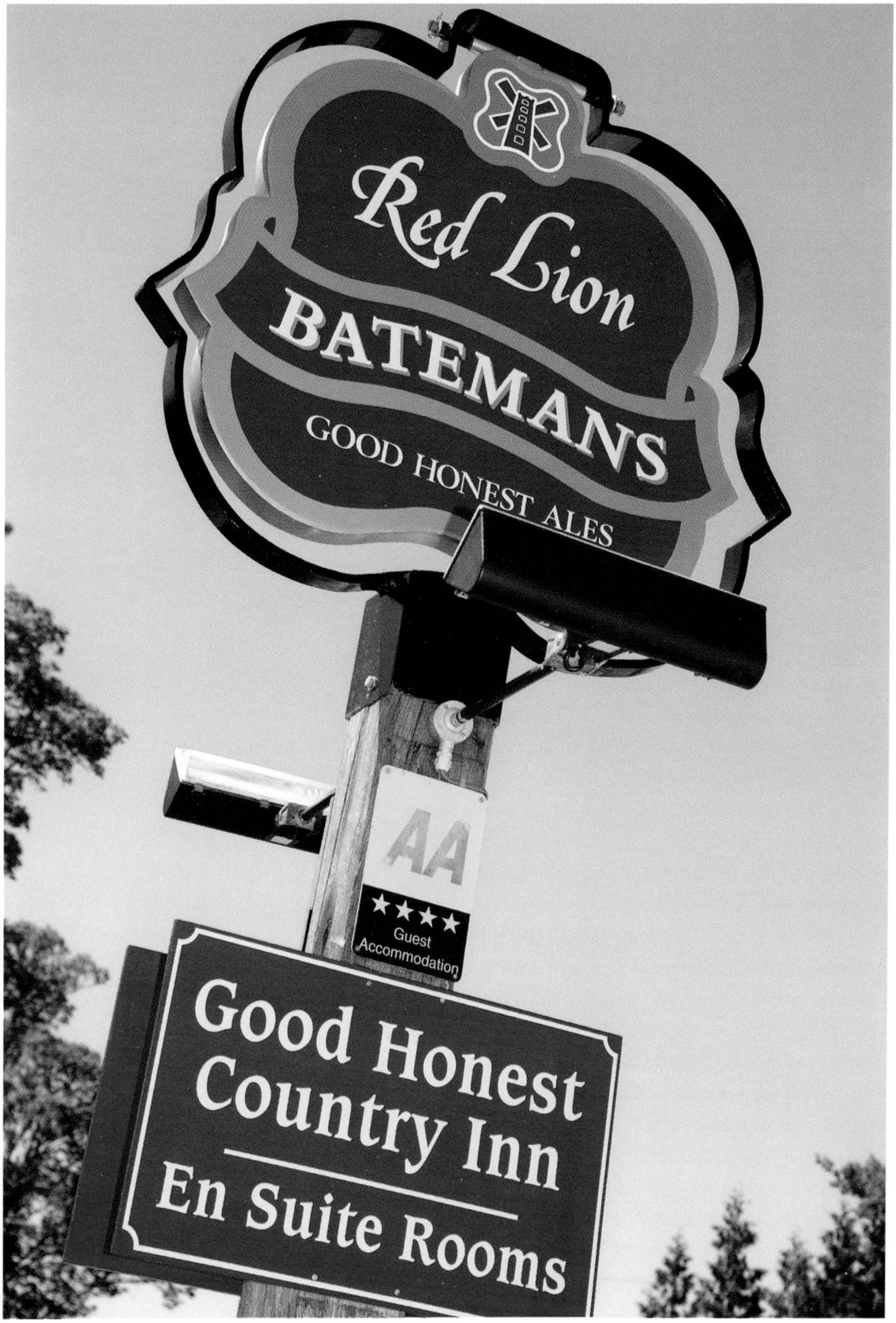

Seeing Red
An example of Batemans' bright and distinctive signage from the mid-2000s at the Red Lion in Revesby. (Batemans)

required to remain closed, and then others when the state of the pandemic allowed some relaxations, such as allowing outdoor trading at pubs. The immediate challenge as the first lockdown descended was stock, both 'wet' and 'dry'. Food from the visitor centre and pubs was sold at half price as takeaways, the proceeds going to charity, and some of the beer stocks were donated to staff and beer club members. In a novel development, Batemans – like some other brewers – recognised that hand-sanitising gel could be produced from alcohol, donating 1,000 litres of XXXB to a distillery for that purpose. All full or unopened stock was returned to the brewery or suppliers, although by the third lockdown most suppliers were refusing to accept returns. Stuart recounts some other problems:

> Part of the frustration in returning beer to the national brewers was that when we subsequently needed stocking up again, one household name sent us back the same stock but had relabelled it, extending the date! The industry as a whole then started to put longer shelf life on their beers. This turned out to be significant, as when the second lockdown was announced our suppliers agreed to once again take back stock which was out of date. More frustrating was when subsequently trade started to really pick up, the national brewers had not forecast sufficient volumes, so we were rationed for stock for a couple of months. First they couldn't brew sufficient beer, then they couldn't get the transport, then they said they were short of kegs, then they said they couldn't get the drivers!

Batemans also wanted to help their tenants directly. Even though publicans were already using government support through business grant schemes (which Stuart describes as 'very good'), they introduced a series of rent concessions, up to 90 per cent dependent on the level of lockdown. The brewery itself was mothballed from 18 March 2020, when Stuart stood on a pallet in the brewery yard and spoke to all the staff about how they needed to go home and look after their families in those troubling times and that everything possible would be done to try and keep the brewery going so that they had jobs to return to. Stuart explained that the company would be put into three phases, the first one being 'Survival', which would be followed by the 'Revival' phase and then ultimately 'Development'. Seventy-seven people – nearly all the employees – were placed on furlough, with 80 per cent of their normal salaries paid by the government. Batemans ensured they kept in touch with all their staff, current and retired, but worse was to follow as the pandemic tightened its grip. Economies had been made wherever possible, but by August the survival of the brewery was once again in question. Batemans made the difficult decision to make twenty-six of their eighty-six staff redundant, as well as closing the reception and visitor centre. Stuart recalls that very difficult and ill-omened day:

> I spoke face to face with all the affected staff. It was extremely sad, but as George Bateman used to say 'sometimes you have to chop off an arm to save the body.' On the very same day the Virginia Creeper which grew around the mill tower, planted by Harry Bateman around 100 years ago, suddenly died – it had cracked in two along its stem. Some of the staff were very superstitious – some of the old country ways still persist – so I bought a replacement plant to represent regrowth and a new beginning.

The rest of the staff remained on furlough, apart from nine who carried on working, mostly on a part-time basis, to ensure there would be a business to come back to when normal times returned. Such measures became very common across the hospitality sector. At least some limited brewing was still needed as there was an increased demand for bottled and take-home beer, particularly through supermarkets and Batemans' online shop, which offered a 'click and collect' service, although bottling had to be temporarily moved from Marston's to Holden's of Dudley. For several months head brewer Scott Lawrence (who'd taken over following Martin Cullimore's retirement) undertook all the brewing and packaging until the other brewers, Carl Leverton and Joey Leeman, who between them had seventy-three years of brewing experience at Batemans, could be taken back on. To meet

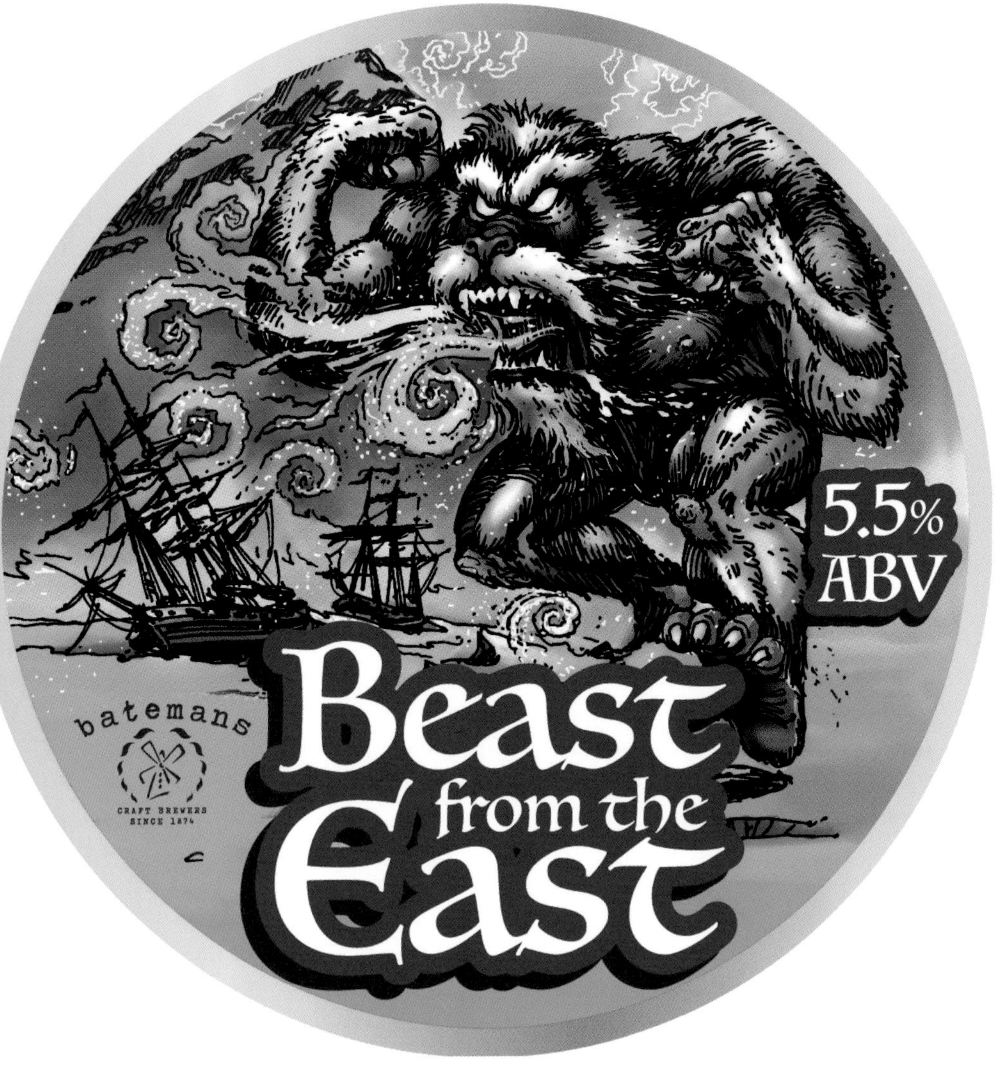

Absolute Beast
The arresting artwork for Beast from the East strong ale. (Batemans)

the demand for take-home trade, canned beers were introduced, starting with Skull and Hammers, an American pale ale, which was followed by Beast from the East, a fearsome canned tipple named after a recent storm that had entered popular folklore. The Beast proved so popular that after the end of lockdown he began to terrorise pubs in cask form. 5-litre mini-casks had become popular across the sector. Batemans introduced them too, but for real ale only with a limited shelf life – they felt quality was important.

Revamped Range
The six permanent bottled beers. Skull and Hammers and Dark Fruits Porter are new beers. The other four received new branding in 2019, replacing the older 'wavy' labels with a more modern square design featuring innovative patterns and appealing typefaces, which stand out well. Rosey Nosey (not seen here) also makes a seasonal appearance later in the year. (Batemans)

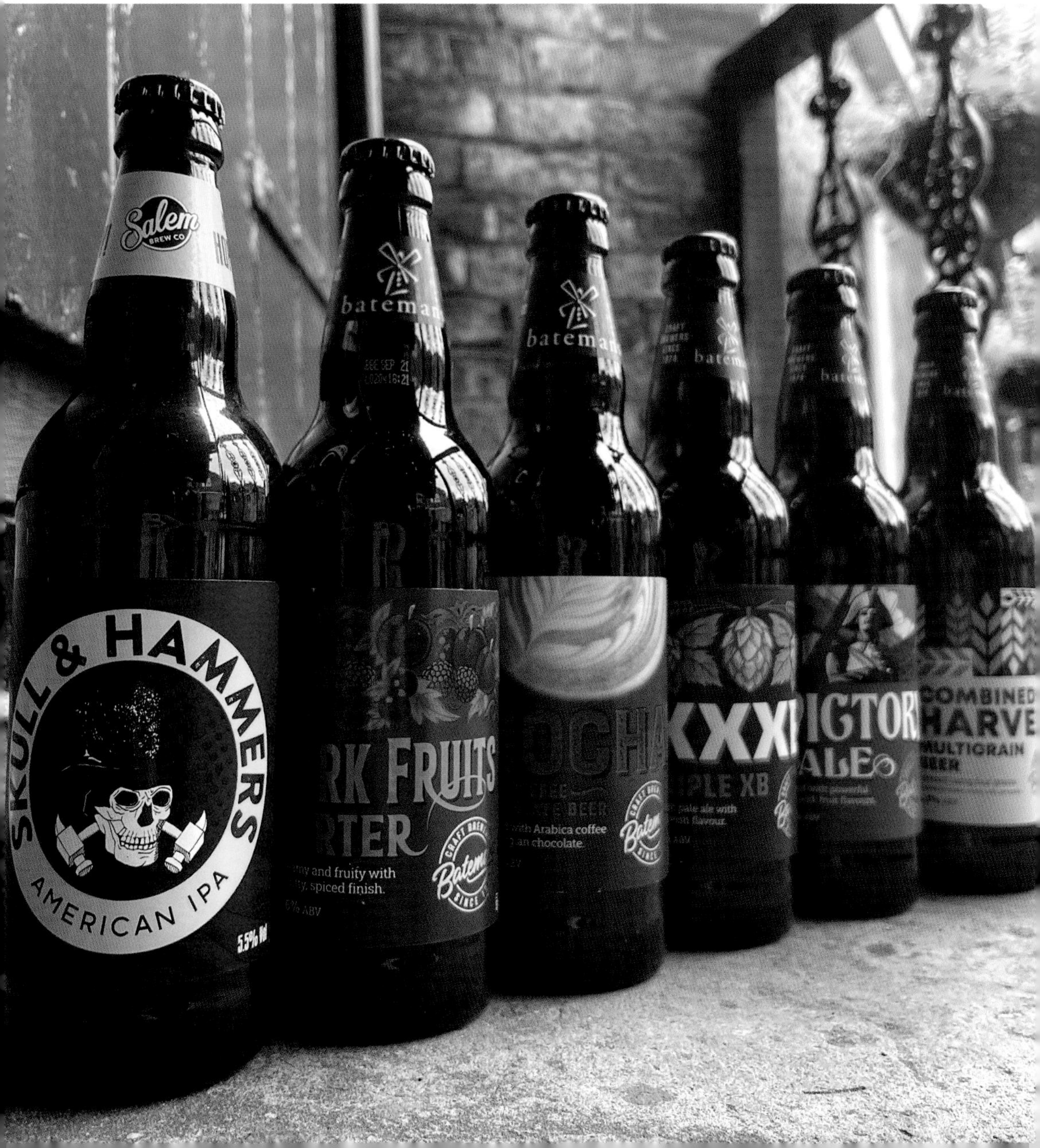

The pandemic even affected the type of beer Batemans could brew. At the onset of lockdown, they'd put their yeast strain into refrigeration. When breweries were given ten days' notice that pubs could reopen, Scott Lawrence felt the culture needed to be built up to full strength. Whilst that took place, other yeast was brought in. Plans to brew an initial batch of XB had to be shelved though: much of the distinctive flavour of a beer comes from the yeast, and it just wouldn't have tasted right if they'd used brought-in yeast. The clever solution was to devise a stopgap beer, called Revival, which allowed XB, XXXB and the golden beer Yella Belly to return a few months later once the original Batemans yeast – which dates back to the earliest years of the brewery – had fully recovered.

Stuart worked continually throughout the Covid period, leading a small team. Sterling efforts had not only been made by the small brewing team, but also by Sue Jacobs, who held the reins in the Operations department, Helen Sales in Accounts, Neil Hutchinson in IT and Chris Jones, who was a constant support to publicans. Slowly, Batemans managed to unfurlough the staff who remained. Between June and October 2020 the brewhouse, telesales and Area Sales Managers returned from furlough, initially part time, as demand began to grow again. In September, sales had almost recovered to pre-Covid levels. By the start of June 2021 all the full-time workers had been taken back in some form, marking a major milestone, although Stuart greatly regretted the departure of Scott Lawrence, who'd moved on to another job. Stepping into his shoes temporarily was the brewery's chief accountant Chris Fuller, who luckily had been studying for his brewing exams, until Buster Grant arrived as the permanent replacement early in 2023. Unusually for a head brewer, Buster has a degree in archaeology but also has a masters in brewing and distillery from Heriot Watt University in Edinburgh. He started his career at Brakspear's brewery in Henley-on-Thames and moved on to South Wales, where he founded the Brecon Brewing microbrewery at Bridgend in south Wales, later moving on to the Cold Black Label brewery in the same town.

Pubs slowly started to reopen, still with Covid precautions and booking-in systems; by July 2021 pub rents returned to normal levels. More decisions were taken to generate at least some income: the Cross Roads pub at East Barkwith in the Wolds was sold as a free house, as well as about 1,500 Bateman casks, which went for scrap. A few non-pub properties were converted into holiday lets, including the former head brewer's house, Merrilodge (complete with swimming pool!), and one of the brewery cottages.

Post-Covid recovery remains challenging right across the hospitality sector. Swathes of pubs and some breweries have never reopened, and others have been unable to recover from major trading challenges, with some high-profile casualties along the way. Batemans have managed to keep such problems at bay, and if you go into one of their pubs these days, you will still find familiar names on the bar: XB and XXXB, with Victory more likely to be found in bottled form. The recipes for XB and XXXB have been very subtly tweaked over the years, with the crystal malt content doubled to 10%, which darkens the beer a little. American Liberty hops have also joined the original Kentish Goldings, this double hopping taking place in the copper, resulting in a new balance of bitterness and aroma. Yella Belly, Skull & Hammers, Salem Porter and its offspring, Dark Fruits Porter (also available in bottles) are also popular. The festive flavours of the Christmas ale, Rosey Nosey, are complemented by the Santa pump clip with its red flashing nose, cotton wool beard and metal bells. The annual ritual to secure enough bottles of Rosey Nosey before it sells out has tasked this writer for many years. But you'll also find one or two seasonal ales, which normally appear for a month at a time. The range is enormous, and includes an old favourite, Dark Mild; although a three times winner of CAMRA's Champion Mild, that was now a generation ago. Mild is no longer a good seller, which has relegated DM to an occasional brew. Batemans ales are also still available in free houses – you will even find their beer (normally XXXB) on board P&O cruise ships. The old barley wine, BBB, made a temporary reappearance in another guise in 2012. Last brewed in the 1970s, a small collection of nips was discovered in a forgotten dark corner of the brewery some forty years later. They had

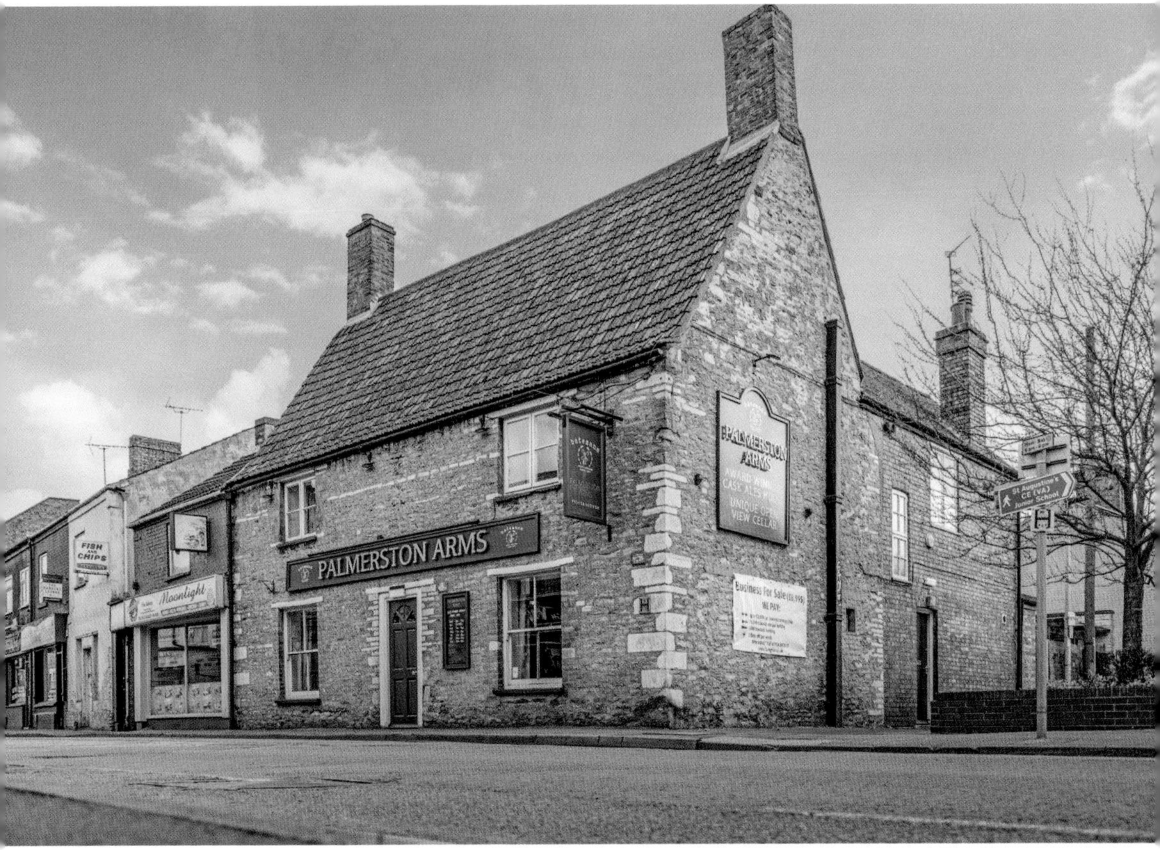

Above and overleaf: **Nice Paint Job**
Recent refurbishments and acquisitions have seen a more subtle approach to branding, especially outside Lincolnshire. The Palmerston Arms, on Oundle Road in Peterborough, has a deep blue colour scheme, while the Champion in Norwich city centre has recently been repainted in a grey palette; both previously had primarily red and yellow signage. A particularly commendable feature at the Champion has been the restoration of the original pub sign, painted on the curved wall, as a reminder of a previous owner, Lacons of Great Yarmouth. (Batemans; Chris Pyrah)

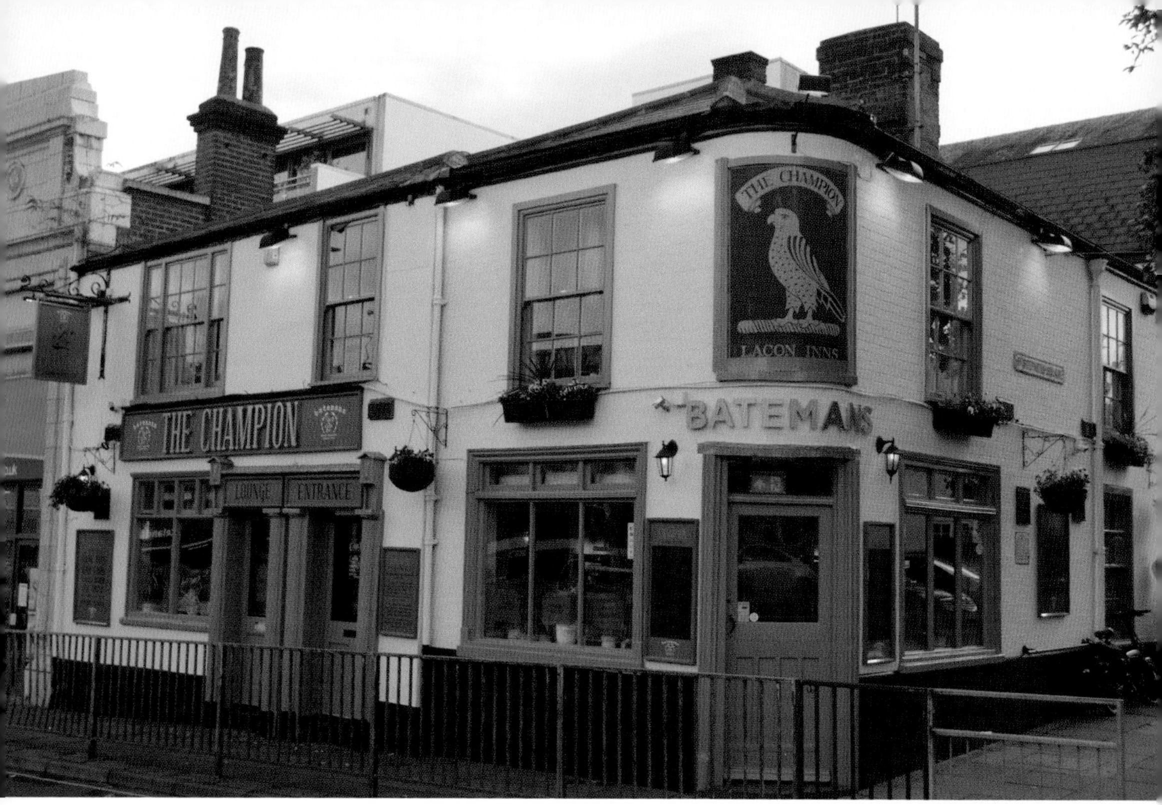

been laid down by George to be opened on Stuart's twenty-first birthday and forgotten about. When opened they were found to be highly drinkable. The recipe was recreated as Vintage Ale for a limited period (best before 2037, mind you, according to the bottles I have!) and I can confirm it certainly is very nice.

Pub acquisitions continue: the seventeenth-century Sanctuary Inn at Bolsover, near Chesterfield, and the lease of the Plough Inn at Horbling, a pretty village just south of Sleaford, were bought in 2023. The Sanctuary (originally the Anchor Inn) operated as a bar, restaurant and B&B, but was closed late in 2022; Batemans rescued it six months later. The Plough was the last pub in Lincolnshire to be owned by a local parish council; there has been a pub on the site for over 200 years. A programme of pub refurbishments is underway too, starting with major investment at the Victoria in uphill Lincoln.

Although the Ridlingtons wine and spirits business in Boston closed some years ago, the brand name has been retained by Jaclyn for two types of gin (distilled by Pin Gin at Louth) and seven wines. The company makes the very best of its placid riverside location, offering caravan and campsite accommodation, and the annual Bands on the Bank music festival, run by Jaclyn. They also offer conferences, wedding venues, a Beer Club loyalty scheme and the ever-popular brewery tour. Despite the closure of the visitor centre as a pub following the onset of Covid, the site is still used to host conferences, wedding events, a Beer Club loyalty scheme and the ever-popular brewery tour. The on-site shop, which was housed in the centre, has been re-housed in the company's offices.

For the future, the fifth generation is waiting in the wings. Having obtained business studies degrees at university, both of Stuart's sons went into the brewery trade, Harry being a production brewer with Beavertown in London, and Edward joining a graduate training scheme at Robinsons in Stockport, having completed an extremely useful year at J. W. Lees Brewery. And daughter Eliza, with a degree in business and fashion, has undertaken some project work at the brewery. But neither Stuart nor Jaclyn show any signs of retirement yet – they are still fizzing with enthusiasm and new ideas, just like their forefathers. Commuting isn't really a problem for Stuart as it takes him an hour and twenty-three minutes to get from home to the brewery ... for the year, and that's only if he walks slowly.

Batemans' independence remains as important as ever in the early twenty-first century as this remarkable brewery, family and staff celebrate 150 years in business. Significant celebrations are planned, as well as a stylish rebrand, and a brand-new beer, 5G, a session IPA whose name acknowledges the five generations who have been key to the story of Batemans.

Here's to the next 150 years!

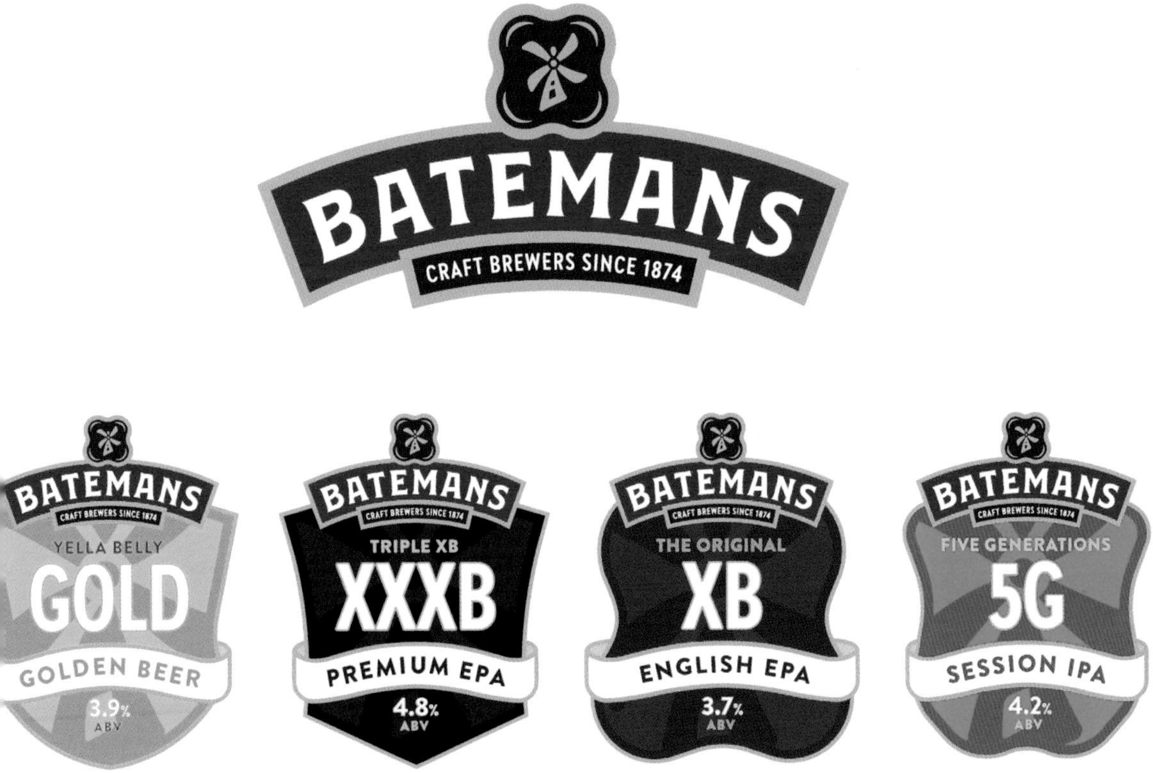

Strong Rebrand
The distinctive new Batemans logo for 2024, together with the new pump clips for four key beers, including the new Five Generations (5G) session IPA. As well as strong consistency across the range, heritage has not been forgotten either: XB and 5G both retain the traditional wavy shape and all four incorporate the refreshed windmill logo, which is echoed in the choice of shading. (Batemans)

Above: Brand Awareness
How the Batemans logo has evolved, starting at the bottom with the original wavy logo from the 1960s, then the 1995 branding update, followed by a refresh in 2014 and finally the new logo for 2024. Note the replacement of 'Good Honest Ales' with 'Craft Brewers Since 1874'. (Batemans)

Opposite: Seagull's View
The brewery complex from above, looking north. In the centre is the mill and visitor centre; to its right are the offices; left and just behind the mill is the original brewery and to its left is the new brewhouse of 2002, emitting steam to show us brewing is underway. In the distance, thunderclouds are brewing over the Wolds. Fortunately that's the only type of storm to affect the firm these days. (Batemans)

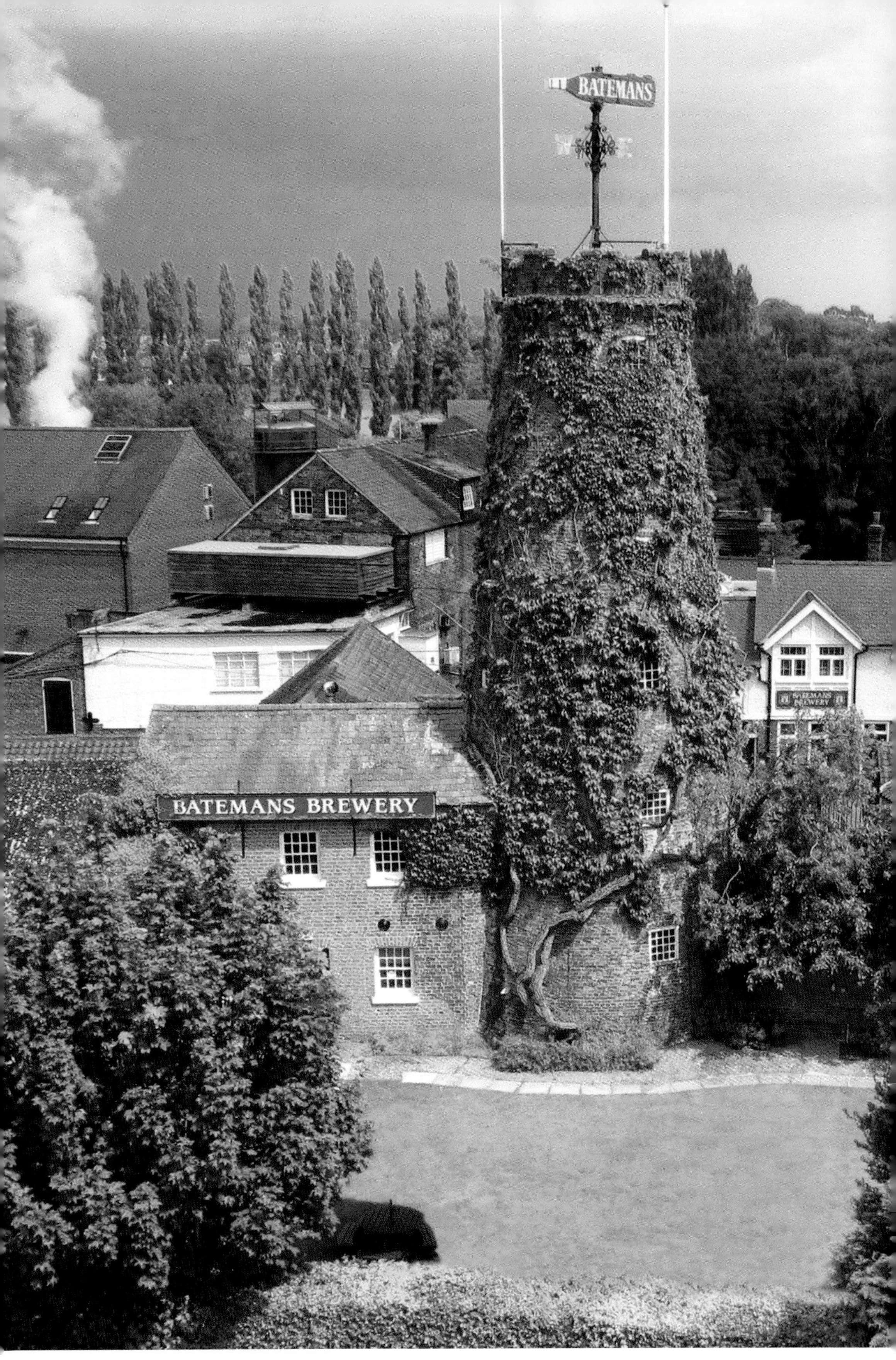
BATEMANS
BATEMANS BREWERY

Appendix 1

Head Brewers at Batemans

Note: Tom Parkes was probably the first to be known as 'head brewer'; his predecessors were probably just known as 'brewer' and there may have been others who were just as involved in brewing.

George Bateman, to *c.* 1910
Ernie Riggall, *c.* 1910–16
Harry Bateman, 1916–19
Cranmer Tom Parkes, 1919–26
James Montague Broad, 1926–52
George Gunson Bateman, 1952–60
Dennis Hartley, 1960–75
Ken Dixon, 1976–84
Stuart Bateman (temporary), 1985
Martin Cullimore, 1985–2017
Scott Lawrence, 2017–21
Chris Fuller (temporary), 2021–23
Buster Grant, 2023–present

Appendix 2

'Independence' (Raise High a Glass)

A poem dedicated to George Gunson Bateman, written and read by his son Stuart at his memorial service held at St Botolph's Church, Boston (the Stump).

A true English gentleman, courageous and brave
Dedication, commitment and loyalty he gave.

He wouldn't sell out and neither would I
Independence, Independence, his battle cry.

His physical presence may be recently gone
But his legacy we promise to carry it on.

Our Brewery will prosper, it carries his name
But without him beside us it won't quite be the same.

He preciously guarded his Brewery and brews
Independence he stated we never should lose.

And long may his beers and milds prevail
To be worthy of the name of Good Honest Ale.

To our Independence George taught us to clasp
And for that Mr George we raise high a glass.

He had countless requests to sell Brewery and beer
But all were sent packing with a flea in their ear.

And now that he's gone maybe they'll come again
They shouldn't waste their breath; it will all be in vain.

Never in life had the dollar been king
Fulfilling work for many was always his thing.

If business was just all about filling the coffers
Then he'd 've stopped years ago boiling beer in his coppers.

And ceased mashing grist in his Victorian mash tun
cos business he said was supposed to be fun.

He was never a lounge man, more public bar
Which is why he's admired by brewers from afar.

But his Brewery came second to tenants and staff
And for that Mr George we raise high a glass.

Death or Glory was his regiment's cry
A value he followed all through his life.

He fought through the '70s with conviction and passion
When it seemed all was lost to the keg beer fashion.

He fought through the '80s when it seemed all was lost
To retain Independence at whatever the cost.

And then in the '90s for his health he did fight
And confounded the doctors and surgeons alike.

And since Year 2000 he battled on still
Kept going by courage, determination and will.

But for all of his suffering, hardship and pain
Never once in that time did I hear him complain.

Independence, Independence, that was all that he asked
And for that Mr George we raise high a glass.

Mischievous and naughty with twinkling eyes
Always the one to shock and surprise.

Played hooker for Boston and they said he was class
Said it was down to him having a rather large *bottom*.

A cheeky boy Charlie was always my Father
More at home in the brew house than office he'd rather.

But forget the Brewery and forget the beer
He was a man with great heart and a man of great cheer.

He overcame unbelievable odds
In his drive to secure local people their jobs.

His brews were all varied, some dark and some pale
But all of them worthy to be good honest ale.

If Dad was a beer he'd be a quart not a half
And for that Mr George, we raise high a glass.

I'm sure God is happy to have him up there
As long as Dad doesn't say 'hey, that's my chair'.

A god-fearing man who believed in what's right
And then tried to achieve it with all of his might.

God will judge him a man full of wisdom and flair
Who developed a culture of kindness and care.

Good honest a phrase that sums the man up
May I take on his mantle and drink from his cup.

And now he looks down in 2007
From the public bar in a corner of Heaven.

And pondering why it was wine changed from water
When there was a choice of bitter, mild or Porter.

But now the flag flies at the Brewery half mast
But for the memory of you Dad, we raise high a glass.

Appendix 3

Covid Poems

During the times of Covid, Stuart kept in touch with staff and customers on a regular basis and sent out two poems to try and keep morale up.

The first was about the brewery windmill and the many ups and downs it had seen during its history.

'The Brewery Windmill'

For 200 years I've looked down from this view
And I'll be here next year – and for years after too
I was here at the time of the Normandy Landing
Even Hitler himself couldn't stop me from standing

I was here when they invented the TV and phone
I was here when Victoria was sat in the throne
And here I still sit at the end of Mill lane
Looking forward to the barrels of beer brewed again

I was here before aeroplanes, space rockets and cars
And I'll still be right here when a man lands on Mars
Many who worked here fought for the cause
They marched right before me as they went off to the wars

The great walls of Babylon eventually fell
But I'm still stood standing and doing quite well.
The Great Depression and two awful world wars
Didn't stop me from opening my doors

I hosted the Dambusters' anniversary do
No wonder Churchill called them 'the few'
They left an impression of how to be behave
In the face of adversity how to be brave

My only companions the pigeons and crows
They've been here beside through the highs and the lows
The steps to my summit are 100 and 2
But remember – the harder the climb the better the view

I'm covered in ivy and Virginia creeper
Not ready am I for the brewery's grim reaper
Though the shutting of pubs wasn't part of the plan
But cometh the hour and cometh the man

It was said I was rotting so they took off my sails
That didn't stop me from brewing my good honest ales
Virus corona for me holds no fear
Cos I'll still be here brewing good honest beer

I've hosted many a party, and board meetings too
And my bar's where you sample any new brew
But by far the greatest party of all
Will be when you return standing proud standing tall

So as windmills go, I've seen quite a lot
From Presidents Lincoln to Kennedy shot
But I'll be here, when the virus is gone
Tall and 'still standing' ask Elton John

I look down on the river and out on the paddock
Across to the chip shop and a nice piece of haddock
But as I stand tall looking far out to sea
I'm wondering what the future might be

Up on my roof flies the Union Jack
Not to wave you goodbye but to welcome you back
The flag will keep flying and I'll still be here
To help symbolise all that's good in our beer

I'm here as a bastion to drive and inspire.
I'm here to see the Phoenix flies out from the fire
I'm here as the guardian of bitter and porter
And I'm here to be a bridge over the troubled water

I'm a symbol for you of endeavour and courage
And look forward to the day when once more we will flourish
Until then when we drink both the grape and the grain
I look forward to the day when we all meet again

The other poem was written in the second lockdown. Its theme was the flag continuing to fly on the brewery windmill, to demonstrate resilience and stubbornness in the face of adversity.

'On My Roof the Flag Still Flies'

Silent streets and silent lanes
Too much beer put down the drains,
Takeaways only, burger and fries
But on my roof the flag still flies

Empty vats and empty pubs
Silent lorries, empty clubs,
Rolling clouds and stormy skies
But on my roof the flag still flies

Part time work and part time furlough
Work from home with a bottle of Merlot,
We'll fight on till the virus dies
But on my roof the flag still flies

Don't hug Gramps, don't kiss Granny
Don't hold hands with Auntie Fanny,
Lonely people, lonely lives
For all of them the flag still flies

Social Distance – Track and Trace
Don't shake hands, don't touch your face,
No more parties, no more pies
But on my roof the flag still flies

Don't go to the pub, for a drink with ya mate
Stay at your home – self isolate,
And travelling far we're told's unwise
But on my roof the flag still flies

One day pubs will open, that is for sure,
Foaming pints of ale once more
'Survival', 'Revival', remains our war-cry
Until that day the flag still flies

But this isn't a war with the German Kaiser
Just wear a mask and pull down your visor,
To honour those who've lost their lives
On my roof the flag still flies

We're still standing defiant as ever
Battered and bruised but beaten? – Never,
Out from the fire the Phoenix will rise
But on my roof the flag still flies

Appendix 4

The Pubs of George Bateman & Son

Most of this information is taken from Batemans' property and mortgage registers, but in some cases exact purchase or sale dates have not been found. Only houses that Batemans owned – rather than supplied – are listed. Pubs are in Lincolnshire unless stated otherwise. Those that are still owned are marked 'C'.

Pub and Location	Bought	Sold
Admiral Nelson, Main Road, Benington	1957	2015
Alexandra Hotel, Siddalls Road, Derby	1989	1991
Anchor, Church End, Friskney	1897	1991
Angel, Church End, Wrangle	1924	1988
Axe & Cleaver, Wainfleet Bank	1918	1966
Axe & Cleaver, Sutton Road, Huttoft	1970	Sold
Axe & Handsaw, New Road, Quadring Eaudike	1951	1972
Ball House, Wainfleet Road, Fishtoft	1957	C
Barkham Arms, Wainfleet Bank	1919	1976
Barley Mow, Main Road, Friskney	1912	1990
Bay Horse, Wincolmlee, Hull	1989	2017
Bell Hotel, High Street, Burgh le Marsh	1921?	C
Bell Inn, Main Road, Halton Holgate	by 1949	1987
Black Bull, Rectory Road, Ruskington	1990	C
Black Bull, Wrangle Bank	1955	1960
Black Horse, Moat Lane, Old Bolingbroke	1921	1985
Black Horse, Siltside, Gosberton Risegate	1931	1984
Black Horse, Thorpe Bank	1920s	1937

Pub and Location	Bought	Sold
Black Swan, South Street, Horncastle	1958	1992
Blacksmiths Arms, Main Road, Skendleby	1926	1988
Blue Bell, Wrangle Bank	by 1933	1975
Boar's Head, Newmarket, Louth	*c.* 1945	Sold
Bricklayers Arms, Friskney Eaudyke	1920	1958
Bricklayers Arms, Wainfleet Road, Old Leake	1924	C
Britannia, Church Street, Boston	1924	C
Bull Hotel, Halton Road, Spilsby	1930	1989
Bull & Dog, Church View, Freiston	1957	1990
Butcher & Beast, High Street, Heighington	1967	C
Butchers Arms, Leake Commonside, Old Leake	1955	1985
Butchers Arms, Brand End Road, Butterwick	1955	1957
Carpenters Arms, Witham Street, Boston	1928	C
Castle Inn, Haltoft End, Freiston	1957	C
Champion, Chapelfield Road, Norwich	2007	C
Chequers, Wainfleet Road, Irby	1949	1954
Coach & Horses, Main Ridge, Boston	1957	C
Coach & Horses, Main Road, Hemingby	1955	1987
Coach & Horses, Main Road, Wrangle	1957	1986
Cottage Inn, Toynton St Peter	1879	1910
County Hotel, North Parade, Skegness	1935	1994
Cross Keys, Victoria Street, Billinghay	1955	1956
Cross Keys, Church Lane, Huttoft	1955	1979
Cross Keys, Queen Street, Horncastle	1933	1978
Cross Keys, Market Place, Belper, Derbyshire	1999	2023
Cross Keys, High Road, Salmonby	1921	by 1943
Cross Roads Inn, Lincoln Road, East Barkwith	2010	2020
Crown, West Street, Horncastle	1970	1987
Crown, Beesby Road, Maltby le Marsh	1955	1990
Crown, Front Street, Trunch, Norfolk	2003	C
Crown, Main Road, Surfleet	1957	1991
Dog & Bone, John Street, Lincoln	1990s	C
Duke of Wellington, Station Road, Midville	by 1948	1976
Duke of Wellington, Hobhole Bank, New Leake	1918	Closed
Duke of York, Lincoln Lane, Boston	1928	2015
Duke of York, Gosberton Risegate	1951	1980
Duke's Head, Duke's Head Lane, Old Bolingbroke	1921	1956

Pub and Location	Bought	Sold
Durham Ox, Main Road, Thimbleby	1933	1970
Ebrington Arms, Main Street, Kirkby on Bain	1955	1957
Fairways, Bawtry Road, Brinsworth, South Yorkshire	2016	C
Ferry Boat, Dogdyke	1957	1957
Fighting Cocks, Market Place, Corby Glen	2000s	C
Five Bells, Church Road, Butterwick	1957	C
Flying Horse, Main Road, Freiston	1957	1960
Foresters Arms, Main Street, Mareham le Fen	1953	*c.* 2000
Fox & Hounds, Main Road, Hundleby	1927	1964
Game Bird, Home Church Lane, Beverley, East Yorks	2014	C
Gate Inn, Bluestone Heath Road, Ulceby	by 1948	1989
George Inn, Lowes Hill, Ripley, Derbyshire	1987	1992
Globe Inn, Fen Road, Stickford	1955	1981
Golden Cross, Church Street, Billinghay	1955	1988
Golden Eagle, High Street, Lincoln	1978	1992
Golden Fleece, Main Road, Wigtoft	1959	2014
Golden Lion, Fishtoft Road, Fishtoft	1955	1991
Good Intent, Church Road, Friskney	1928	1969
Great Northern, Station Road, Surfleet	1913	1987
Green Man, Old Main Road, Scamblesby	1970	1988
Greyhound, Sea Dyke Way, Marshchapel	1984	1991
Half Moon, West Street, Alford	1970	1986
Horseshoes, Main Street, Asfordby, Leics	2015	2023
Indian Queen, Dolphin Lane, Boston	1970	2016
The Ivy, Market Place, Wragby	*c.* 2000	2009
Jolly Sailor, St John Street, Wainfleet	1992	2009
King William IV, Horncastle Road, Boston	1957	2004
King's Arms, Hall Road, New Lakenham, Norfolk	2004	C
King's Arms, Horncastle Road, Boston	1957	C
King's Head, Yeoman Street, Bonsall, Derbyshire	2002	C
Kings Head, Wainfleet Road, Firsby	1928	1959
King's Head, Church Road, Freiston	1957	C
King's Head, Bull Ring, Horncastle	1923	C
Ladybower Inn, Bamford, Derbyshire	2015	C
Lancaster Arms, Main Road, New Bolingbroke	1955	1956
Leopard, Bull Close Road, Norwich	2007	2014
Lincolnshire Poacher, Mansfield Road, Nottingham	1987	Sold

Pub and Location	Bought	Sold
Lion Inn, Mosley Street, Basford, Notts	2001	Sold
Londesborough Arms, Middle Street, Metheringham	1954	C
Lord John Russell, Russell Street, St Neots, Cambs	2000s	2016
Lord Nelson, Trafalgar Road West, Gorleston, Norfolk	2009	C
Lord Raglan, Foundry Street, Horncastle	1929	1985
Malt Shovel, Northgate, Louth	1984	1987
Midge Inn, Main Road, Hatton	1930s	1978
Mill Inn, West Street, Billinghay	1955	1978
Mill Inn, Spilsby Road, Boston	1957	C
Nags Head, Foundry Street, Horncastle	1929	Sold
Nags Head, Southgate, Sleaford	1984	2017
Nags Head, Kirton Holme Road, Kirton Holme	1955	1968
Napoleon, Fishtoft Road, Boston	1988	2017
Nelson Butt, Market Street, Spilsby	1932	Sold
Nelson's Head, Horatio Street, Bethnal Green, London	1989	1990s
New Castle, Fydell Street, Boston	1970	2011
New Chequers, Croft	by 1948	1957
New Inn, High Street, Wainfleet	1920	1988
New Inn, Fodder Dyke, Friskney	1957	1957
New Inn, Fosdyke Bridge	1957	1957
New Park, South End, Boston	1988	2000s
Oatsheaf, Hobhole Bank, New Leake	1922	1956
Odd House, Fletcher Street, Grantham	1988	c2000
Old Bridge Hotel, Market Street, Holmfirth, West Yorks	2003	2015
Old Chequers, Lymn Bank, Croft	1904	C
Olive Bar & Grill, Station Road, Barnby Dun, South Yorks	2015	C
Ostrich Inn, Moathouse Lane, Leverton	1957	1970
Packet Inn, Witham Bank, Dogdyke	1926	Sold
Packet Inn, Moorhouses Road, Tumby	by 1933	Sold
Palmerston Arms, Oundle Road, Woodston, Peterborough	2000s	C
Parade Hotel, Grand Parade, Skegness	1972	1991
Peacock & Royal, Market Place, Boston	1926	1965
Peacock Tap, Market Place, Boston	1926	1965
Pied Bull, Hilldyke, Sibsey	1957	1978
Plough, The Green, Nettleham	1931	C
Plough, Station Road, Swineshead	1954	1984
Plough, Fen Road, Halton Fenside	1918	1960

Pub and Location	Bought	Sold
Plough, Main Road, Covenham	2000s	2011
Plough, Main Road, Stickney	1955	1966
Plough, East Street, Horncastle	1955	1956
Plough, Sea Lane, Wrangle	1955	1955
Plough, Spring Lane, Horbling	2023	C
Prince Albert, Irby	1928	1959
Princess Victoria, South Parade, Matlock Bath	2000s	2018
Queens Head, Boston Road, Spilsby	1923	1987
Queens Head, Station Road, Legbourne	1922	1993
Railway Inn, Aby	1955	1990
Railway Hotel, Station Road, Firsby	1918	1972
Railway Hotel, Witham Road, Kirkstead	1933	2014
Railway Inn, Sleaford Road, Tattershall	by 1927	Sold
Railway Inn, Southrey	1955	1957
Railway Inn, Station Street, Mansfield, Notts	by 2002	Sold
Red Cow, Gaysfield Road, Fishtoft	1957	C
Red Lion, Hogsthorpe Road, Mumby	1955	C
Red Lion, Burgh Road, Orby	1955	1985
Red Lion, Main Road, Great Steeping	1957	1957
Red Lion, High Street, Wainfleet	1906	2013
Red Lion, Market Street, Spilsby	1954	1991
Red Lion, East End, Burgh le Marsh	1920	1988
Red Lion, Main Road, Revesby	1974	C
Red Lion, Main Road, Baumber	1969	1992
Red Lion, High Street, Swaffham Prior, Cambs	2009	C
Red Lion, Fen Road, East Kirkby	1931	1988
Rising Sun, Main Road, Stickney	1967	2021
Rising Sun, North Road, Bratoft	1921	1955
Rising Sun, Sausthorpe Road, Hagworthingham	by 1948	1955
Robin Hood & Little John, Main Road, New Bolingbroke	1948	1966
Robin Hood, High Street, Boston	1988	Sold
Ropers Arms, Horncastle Road, Boston	1957	2000s
Rose & Crown, Main Road, Stickney	1957	1986
Royal Oak, Main Road, Mareham le Fen	1957	C
Royal Oak, Main Road, New Bolingbroke	1957	1991
Royal Oak, High Street, Wainfleet	1919	2016
Royal Oak, Roman Bank, Winthorpe	1984	1991

Pub and Location	Bought	Sold
Sailor's Home, Sea Lane, Old Leake	1916	1955
Sanctuary Inn, Market Place, Bolsover, Derbyshire	2023	C
Saracens Head, Main Road, East Keal	1921	1977
Sebastopol, Church Lane, Minting	1951	1987
Shepherd's Arms, Main Road, Wrangle	1904	1957
Ship Inn, Sea Road, Chapel St Leonards	1970	C
Ship Albion, Albion Street, Spalding	1989	1994
Ship Tavern, Custom House Lane, Boston	1918	2018
Ship Inn, Main Road, Fosdyke Bridge	1923	Sold
Ship Inn, Fen Road, Toynton All Saints	1910	1964
Ship Inn, Wharf Road, Retford	2002	C
Ship Inn, Sibsey Northlands	1957	1957
Stags Head, Main Road, Burwell	1984	2001
Station Hotel, Station Street, Holbeach	1925	1977
Three Horse Shoes, Sheepgate, Leverton	1963	1972
Three Tuns, Thurlby Road, Bilsby	1955	1990
Three Tuns, Main Road, Ingoldmells	1920	2016
Three Tuns, Wrights Lane, Friskney	1920	1957
Three Tuns, Culvert Road, Thorpe Culvert	1936	1989
Treaty of Commerce, High Street, Lincoln	2001	C
Vanguard, Main Road, Keal Cotes	1979	1992
Victoria Tavern, Tavern Lane, Hogsthorpe	1920	1989
Victoria Inn, Union Road, Lincoln	*c.* 2005	C
Vine, Church Lane, South Thoresby	1930?	2009
Vine Hotel, Vine Road, Skegness	1929	Sold
Vine, Barnby Gate, Newark, Notts	1988	2021
Waggon and Horses, Kirton Holme Road, Kirton Holme	1928	1985
Waggon and Horses, Lawrence Street, York	c2005	C
Wheatsheaf, Spilsby Road, New Leake	1928	1989
Wheel Inn, Hanby Lane, Welton le Marsh	1920	1977
White Hart, South Street, Alford	1970	1989
White Hart, Main Road, Sibsey	1957	C
White Hart, Westville Road, Frithville	1957	1961
White Hart, Church Road, Old Leake	1957	C
White Hart, High Street, Nettleham	1957	C
White Hart, The Highway, Great Staughton, Cambs	*c.* 2005	C
White Horse, Fold Hill, Old Leake	1955	1956

Pub and Location	Bought	Sold
White Lion, Main Road, Little Steeping	1913	1979
White Swan, Old Boston Road, Coningsby	1924	2006
White Swan, High Street, Burgh le Marsh	1920	1991
White Swan, Scotgate, Stamford	1988	Sold
Willoughby Arms, Church Lane, Willoughby	1977	C
Windmill, Silver Street, Coningsby	1957	1987
Windmill, Spilsby Road, New Leake	1922	1957
Windmill, Chapel Road, Old Leake Commonside	1914	1966
Woolpack, Riverhead Road, Louth	1925	C
Woolpack, High Street, Wainfleet	1989	C
Wormgate Inn, Wormgate, Boston	1988	1990

Appendix 5

Tales, Facts and Confessions

It's nearly closing time now, but there's just space for an eclectic selection of short tales and anecdotes, some interesting, some amusing and some revealing! Whether you dip in and out of them, or read them all at one go, either way you'll get a good sense of what it means to be an independent brewing company in the early twenty-first century, and some of the high points along the way.

Brewing

- A beer and brewing presentation was carried out by a sales manager at the Trent Bridge Inn in Nottingham. As its focus was on yeast, he had brought some samples in a glass jar. Afterwards, he went to a restaurant and put the case of samples under the table. All of a sudden pandemonium broke out, with the customers diving under their tables – the yeast jar had exploded, and had been mistaken for a bomb.
- Before effective thermometers, prior to 'mashing in' the brewer would judge the temperature of the hot liquor by how well he could see his reflection on the surface of the water. If it was too hot, there was lots of steam and he couldn't make out his reflection.
- The old bottling machine, which was decommissioned in the early 1990s, had been bought from a scrapyard, having been used for bottling soft drinks. The original pasteuriser was bought from a local farmer who had used it for pasteurising milk.
- When he was the brewer in the 1950s and early 1960s, George Bateman would sometimes undertake the early brew in his dinner suit with bow tie if he arrived home late from a function.
- George painted the slates on the fermenting room roof white in 1951, to try and reflect the heat from the sun. Early air conditioning!
- Batemans still use the same strain of yeast they used in 1874.
- When the 'goodness' has been extracted in the brewing process from the malted barley grains, Batemans send them off to a local farmer for cattle food.

- Batemans once sent milled malt to a local farmer, which had not been used in the brewing process, as it had been rejected on the grounds of poor quality. The farmer was advised to dilute the malt with cattle feed. He failed to do this and all the cattle blew up as the yeast in their stomachs reacted with the sugar in the malt, producing carbon dioxide.
- It takes 5 pints of water to brew one pint of Batemans beer. It's the intention to get this down to 4, then 3.
- When Batemans had wooden barrels, after they had been washed, the brewery's 'sniffer' would smell the inside of them to ensure they smelt sweet. All the empties waiting to be filled had to be hosed down every day to ensure the reeds in between the staves did not dry out.
- One of Batemans' warehouses is a Romney hut, which replaced an Iris hut, both prefabricated steel structures used by the British military in the Second World War. It's newer than the mechanics' shed, which dates back to the First World War!
- Batemans were asked by Two Dogs alcoholic lemonade to contract bottle for them. Originally, Batemans accepted the opportunity, but then realised they wanted bottling each day what would normally take three months to package!
- Batemans' present Head Brewer, Buster Grant, got his taste for beer after volunteering to help on the Batemans stand at the Great British Beer Festival.
- The main warehouse is called 'The New Warehouse' because it was new when Batemans built it – that was over forty-five years ago!
- When the new Brewhouse was built in 2002, it wasn't built tall enough and Batemans couldn't fit the equipment in. It had to be taken out and more layers of bricks laid.
- During 2012, five cases of Batemans BBB strong ale (7.5% ABV!) were found in the cellars, where they'd been laid down to celebrate Stuart's twenty-first birthday in 1981 and then forgotten about. They had been there for thirty-five years, by which time Stuart was fifty-two.
- In 1999, young Harry Bateman found thirty-six bottles of Taylor's 1921 port (with no labels on them) in one of the old, deserted cellars. His mum, Rachael, used one of them to make a stew. It turned out some of the bottles were worth £450 each.
- When George Bateman's car expired in 2002 having done 199,973 miles, the scrap man wouldn't take it, so it was raffled off for charity and raised £1,500.

Property

- Batemans had a hotel in Skegness – the Vine – where the skeleton of an exciseman was found bricked up behind a wall.
- When Batemans bought The Woolpack in Wainfleet from Scottish and Newcastle Breweries, rather than pay for it they agreed to sell their McEwan's lager in their pubs. Not an easy sell south of the border!
- Harry Bateman built a hotel on Skegness seafront called the County. Its foundation stone was laid by siblings George, John and Helen, and it still bears their engraved initials. It was outside the same hotel that some years later Stuart

was 'kidnapped' by a visiting rugby team who stripped him off and chained him to the railings.
- Batemans had a pub, the Royal Oak, at Wainfleet, which had once been the local asylum.

People

- Neither Harry Bateman nor George Bateman retired; they both worked full time, up until the last.
- George's father-in-law, Frederick Carson, was a doctor who was awarded the Military Cross in the First World War for pulling injured soldiers out of no-man's land.
- When sleeve (straight) pint glasses were introduced, Batemans' landlords made sure they all had one traditional dimple glass behind the bar in case George Bateman came in. That was his favourite style of glass.
- In 1973 Colin Crowson joined the brewery from Watney Mann, on a one-year contract to arrange the centenary celebrations. He stayed twenty years.
- The longest-serving member of staff was Harold (Harry) Waugh who worked at the brewery for fifty-one years.
- Members of one family, the Symonds, worked at Batemans for a combined total of nearly 220 years.
- In the mid-1980s, 'Happy' Brett, who worked in the cellars, tried to mislead George into believing he was not sixty-five so he didn't have to retire, for fear of losing his beer allowance.
- One Head Brewer crashed his car into a dyke and wasn't found for two days. He had been short on petrol and turned his engine off in order to cruise, not realising he'd also turned the steering lock on, until he came to a corner.
- Both Bateman boys currently work for competitor breweries: Harry at Beavertown in London and Ed at Robinsons of Stockport, having previously worked at John-Willie Lees in Manchester.
- Harry Bateman and Billy Butlin were good mates, with Billy building a bowling alley next to Harry's County Hotel.
- Batemans had a Head Brewer who resigned because he'd decided to go teetotal.
- Batemans had an eccentric surveyor years ago who used to bring a picnic hamper to work each day and would lay out a rug on a grass verge to eat his lunch on it.
- If George Bateman hadn't been a brewer, he would have been a vicar.
- Independent family brewers held a procession outside the Houses of Parliament in 1997 to protest against rises in beer duty. Most of the brewers took their vintage drays, but the Bateman family just took a pram with their son Harry in it.
- Stuart trained in all departments at Mansfield Brewery. However, he was moved from Cellar Services after he'd gone to change a dispense tap at the Aquarius Club in Chesterfield, only to set the trip switches off on all the pumps whilst all the drip trays were out, ending with beer all over the bar floor. This was followed by Stuart changing a pump at Ripley Working Men's Club, which resulted in beer shooting over his shoulder onto the newly polished mirrors.

- When Stuart was training as a pub manager at The Bench and Bar, next to the Broadmarsh Centre in Nottingham, he was surprised that no one came in for the first two hours of his first shift, until he realised he hadn't unlocked the doors!
- Part of Stuart's training was to take his HGV 1 lorry driver's licence when he was twenty-one. On his first trip out on a fully laden lorry he forgot to put the side rails on the trailer. All the barrels tipped off the side as he drove round the first corner and proceeded to bounce down the road, overtaking the lorry.
- Rachael Bateman went out with the family for dinner and invited beer writer Roger Protz. Upon talking about Michael Jackson (another famous beer writer) she innocently said that she didn't know about his writing, but thought he was a great dancer!
- Roxy Beaujolais, the legendary publican of the Seven Stars in Holborn, London, banned Stuart from her pub, as he had innocently eaten the 'picky bits' out of the saucer on the bar, thinking they were snacks. They were actually Roxy's cat food.
- One of the first matches Brian Clough's Nottingham Forest played as a newly promoted top flight team in 1977 was a friendly against Skegness Town, sponsored by Batemans. The only type of payment Brian Clough wanted was a few bottles of champagne.
- At a Chas and Dave concert in Skegness, Chas said that he looked forward very much to returning, as the beer was so good in Lincolnshire.
- When Batemans opened their new Ridlingtons wine warehouse in Boston, a reception was arranged for customers and dignitaries. A member of the Batemans sales team asked if there was a dress code for the evening and was jokingly told, as a wind up, that there was a theme of Old Time Music Hall (which of course there wasn't). Nothing more was said and the conversation was forgotten about, until the evening of the event when he walked into the function, where everyone else was suited and booted, dressed as a First World War soldier wearing a gas mask!
- 29 Squadron, which was based at RAF Coningsby, was nicknamed Triple XB Squadron. A local signwriter was asked to paint a sign with 'XX1X' on it, but he interpreted the instruction as 'two X's and one X', which he took to mean three Xs, and painted 'XXX'.
- One of the last reunions of 617 Dambusters Squadron, Royal Air Force, took place at Batemans in 1993, when they officially opened the visitor centre.
- During discussions with directors from a national brewer, negotiations weren't going too well. Consequently, a Batemans Commercial Manager locked them in the brewery windmill until they saw his way of thinking. With the weekend beckoning, they soon saw sense!
- George took his goat out for a walk one evening, but as he'd been gone for a long time Stuart thought he'd better go and look for him. After an hour of walking round Wainfleet, Stuart gave up and went to The Jolly Sailor, his local, for a pint. There in the corner with a pint was George, with the goat sitting on the next chair.

Trips

- George Bateman went on a trip to Kaltenberg Brewery in Germany, which was also attended by a distinguished member of the Whitbread brewing dynasty. At registration Mr Whitbread was asked for his details, to which he merely stated

'Whitbread of England'. George was up next: 'Bateman of Wainfleet' was his tongue-in-cheek response.

- In the early 1990s Stuart went on an industry trip with the national brewer Ansell's, all the way to Australia to visit Castlemaine Brewery in Brisbane. When they got there it was closed.
- Batemans' sales director went to the Oktoberfest, a German beer festival, where there was a slight problem when one of the party they were with suddenly stood up, having found a chicken leg stuck down the back of his Lederhosen. He proceeded to thump the bloke who was sat behind him, who was in the seat that the sales director had just vacated – having been eating chicken! A major international incident followed.
- One of the highlights of the brewery entertainment calendar is the Heineken ski trip, attended by many from the industry. Stuart was once presented with a three-armed shirt (a reference to the allegedly rather small gene pool in Lincolnshire).
- Batemans held a competition with the national wholesaler The Beer Seller, which revolved around their Jolly Fisherman range of beers. The awards were presented in Paris to an audience of 150 Beer Seller staff, by a brewery director dressed as the Skegness Jolly Fisherman. The outfit involved wearing a massive plastic head which you had to look out through the nostrils to see, and a costume stuffed with six large pillows. When he ran on to the stage he tripped over and couldn't get back on to his feet. The main act, ventriloquist Roger De Courcey, famed for his puppet Nookie Bear, complained that he'd never been so upstaged in the whole of his career.
- At a Britvic sales conference held at a hotel in Malta, Batemans' then sales director went to bed in the early hours after a very good final night. He decided to put his luggage outside his room for it to be collected in the morning. However, as he wrestled his bags out of his bedroom the door clicked shut behind him, and he was left in the corridor wearing no clothes. As it was so late and no one seemed to be about, he thought he'd go down into reception and see if the night porter could help. The lift stopped at the ground floor and the doors opened. There in front of him were twenty French students who had just booked in.
- One of the first regular hospitality trips in the industry was the annual Carlsberg Titlark trip to Copenhagen, named after a canal boat. Upon returning home from the trip one of the Batemans team told his father how one of the brewers had won a trophy for being the 'most entertaining' person on tour. George unexpectedly responded with 'Yes, I won that when I went!'
- A sales group went on a team-building exercise on a boat on the Norfolk Broads. The diesel tank ended up being filled with petrol, the bicycles all fell off the roof into the water and the driver (who shall remain nameless) hit another boat, leaving a hole in it.
- At a brewers' Christmas lunch held at Laine Pub Company's The Aeronaut in West London, Oliver Robinson of Robinson's Brewery surreptitiously arranged a whip round for charity resulting in Stuart having to dress up in the entertainer's sweaty tiger outfit and sing 'Tiger Feet'.
- Three Batemans national account managers were staying in a hotel in the North East, prior to a sales presentation at Federation Brewery. They played 'spoof' into the early hours and having exhausted the contents of the mini-bar as penalties, it was decided

the new penalty was to run out through the French doors and perform one lap of the garden, naked. Unfortunately for the loser the two others locked him out and he had to come back in via reception.

- There was a brilliant industry trip with Miller Brands to Ghent. It included a tour around the local monastery, led by one of the monks. The Batemans contingent tried to get him to come out with them for the evening, but he said he couldn't as he had an exorcism to perform!
- When Jaclyn and her now ex-husband Jim got married, Stuart was best man. He arranged to take Jim on a pre-wedding trip to the Isle of Man to visit pubs and Okell's Brewery. The only instruction Jaclyn gave to Stuart was to bring Jim back. The one thing Stuart didn't do was exactly that!

Sporting Events

- Batemans helped to arrange the first North versus South brewers' cricket match, played in Chiswick, West London. The North didn't cover themselves in glory, being out for less than 20, with their talisman, former Scotland batsman Steve Crawley, then of Caledonian Brewery, out first ball. And he was the only one who had his own bat and got the email about having to wear whites!
- Stuart played rugby at Twickenham in a charity game alongside Ted Tuppen, who was CEO of Enterprise Inns. Ted asked Stuart after the match what he was doing at the weekend. Stuart said that he would be in his local club and would do the meat raffle and if he won he'd send Ted the meat. This led Ted to be somewhat perplexed, particularly when Stuart (who didn't win the raffle, or even enter it!) bought some pork chops and liver, wrapped them up, by which time they were rancid, and sent them to Ted at Enterprise HQ.
- Another great sporting event that Batemans helped to set up was the first North versus South Brewers' football match. However, at the game, played in Burton-on-Trent, the referee threatened to not start the match as the Northern team did not have shin pads. What came to the rescue was an old hamster cage found in a dyke next to the pitch. At the bottom of the cage was some dirty lino which the team ripped into squares and pressed into service as makeshift pads. The referee shrugged his shoulders and agreed to play the game.
- Batemans became renowned for their industry golf days in Skegness, which were called 'Bit of a Bash by the Beach'. The days were famous for players being presented with whatever outlandish shirts could be found to be worn for the event: fluorescent yellow, tie dye, stars and stripes, and even a light grey polo shirt which had been sprayed darker grey under the armpits to make it look sweaty. However, one year, rather than shirts, socks were presented to players which had bells attached to them.
- Carlsberg, who sponsored the old Wembley Stadium, arranged their last football knock-out tournament there just before it was demolished, between several teams from the on trade and off trade. Stuart's team went through all the rounds to the final. The two sides couldn't be separated, even after extra time and five penalties. The opposition

then missed their sixth penalty and finally Stuart put the ball in the top corner to score the last goal at Wembley in a tournament organised by Carlsberg at Wembley.

- Stuart sat next to the famous rugby international Martin Bayfield at a sportsman's lunch. Martin asked Stuart where he was from, to which Stuart shyly responded that he was from a very small brewery which he would never have heard of. Martin persuaded him to tell him the name of the brewery as he loved real ale. When Stuart told him he said 'Batemans Brewery, really?' to which Stuart excitedly responded 'Yes, have you heard of us?' – 'NOPE' was the comment from the big-man to the deflated Stuart!

Export

- George and Stuart were off on an export trip, their car loaded up with beer and luggage. George had to stop off at Southampton before catching the ferry, to collect some goods for a friend which needed taking to Spain. There turned out to be a lot more to collect than was envisaged. Stuart was a bit worried that George would say that they would have to ditch some of the beer. But in true Bateman spirit George's response was 'Well, we'd better get rid of the luggage then!'
- On an export trip to Paris held on a boat on the Seine, Batemans set up a little trestle table next door to the huge Bass exhibition stand. The Bass staff left at lunchtime on the last day, leaving behind half their stock of Bass Export. Stuart spent time trying to distract his father so he could remove the abandoned beer from the stand to his car, but eventually gave up and went for a walk. When he returned, he found that George had managed to get the entire stock of Bass Export into his car boot!
- Jaclyn exhibited Batemans bottled beers at a food and drink exhibition in Sydney, Australia, in the early 1990s. The exhibition started on a Sunday, but there was no beer on the Batemans' stand as the container was still in dock. When the bottles did arrive at the exhibition, ready for the opening on the Monday, all their front labels had been applied upside down. Quite appropriate really!
- Batemans used to import port in barrels from Portugal, to be decanted into bottles.

Film and TV Appearances

- A Batemans beer case featured in *EastEnders*. Phil Mitchell sat on it in the pub cellar for a twenty-second shot.
- The 2007 Nicole Kidman film *Golden Compass* featured a Batemans beer case that had previously appeared in *EastEnders*. In an interview with BBC Radio Lincolnshire, Stuart was asked why it had been selected. He explained that it had been by far the best beer case at the screen test. The interviewer actually asked if that was true!
- Batemans pubs featured in the 2019 ITV detective series *Wild Bill*, which starred Hollywood actor Rob Lowe as a former Miami detective heading up the fictitious East Lincolnshire Police Force.

- In Jeffrey Archer's book *Mightier Than the Sword,* the author refers to nights out in nightclubs in the south of France, but 'where will I find a pint of Batemans and cod and chips served in newspaper?' The popular author knew Lincolnshire well, having been the MP for Louth many years ago, and later served part of his jail sentence for perjury at North Sea Camp and Lincoln prisons.
- Batemans bottles, drip mats and bar towels appear on the bar top during the pub brawl scene in the Krays film *Legend,* which starred Tom Hardy as both of the Kray twins.
- The TV programme *Salvage Hunters* visited the brewery, but only bought some old beer crates for £50, which they subsequently returned as they were damaged.
- Batemans brewed a special beer for Guy Martin's sister's pub, the Marrowbone and Cleaver, in Kirmington, north Lincolnshire, named 'Skull and Spanners Beer' after Guy's logo.

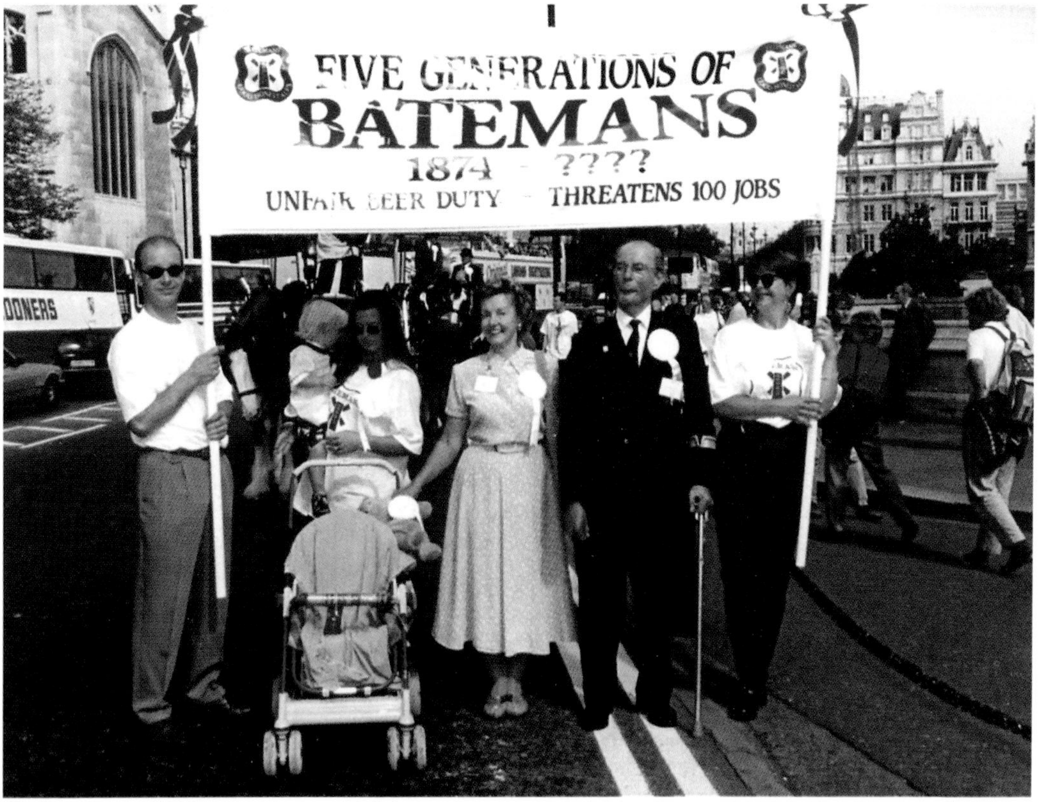

The Bateman family campaigning in London alongside other family brewers, some of whom have now gone out of business.